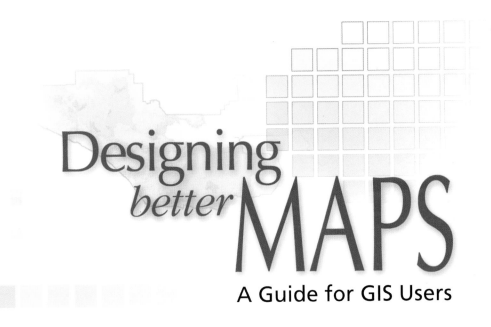

Designing *better* MAPS

A Guide for GIS Users

Cynthia A. Brewer

ESRI PRESS

REDLANDS, CALIFORNIA

Esri Press, 380 New York Street, Redlands, California 92373-8100

All rights reserved. First edition 2005
17 16 15 14 13   5 6 7 8 9 10
Printed in the United States of America

Library of Congress Cataloging-in-Publication Data
Brewer, Cynthia A., 1960-
    Designing better maps : a guide for GIS users / Cynthia A. Brewer.—1st ed.
        p.   cm.
    Includes bibliographical references.
    ISBN 1-58948-089-9 (pbk. : alk. paper)
    1. Cartography. 2. Geographic information systems. I. Title.
    GA105.3B74   2005
    526—dc22   2005007987

ISBN-13: 978-1-58948-089-6
ISBN-10: 1-58948-089-9

Ask for Esri Press titles at your local bookstore or order by calling 800-447-9778, or shop online at www.esri.com/esripress. Outside the United States, contact your local Esri distributor or shop online at www.eurospanbookstore.com/Esri.

Esri Press titles are distributed to the trade by the following:

In North America:
Ingram Publisher Services
Toll-free telephone: (800) 648-3104
Toll-free fax: (800) 838-1149
E-mail: customerservice@ingrampublisherservices.com

In the United Kingdom, Europe, Middle East and Africa, Asia, and Australia:
Eurospan Group
3 Henrietta Street
London WC2E 8LU
United Kingdom
Telephone: 44(0) 1767 604972
Fax: 44(0) 1767 601640
E-mail: eurospan@turpin-distribution.com

CONTENTS

CONTENTS, continued

CONTENTS, continued

PREFACE

I have been teaching map design for twenty years as a graduate student and professor. The guidance I have organized for this book combines what I have learned from my professors, supervisors, and students over those years. I have also learned map design by doing research on map reading and by making maps professionally. My research and consulting work with people making maps at varied federal agencies taught me that GIS and statistical experts who do not have cartographic training are seeking practical advice on how to design their maps. This book is intended as a basic guide for people who want to improve the maps they make.

Map design principles have been fairly stable over time, but the tools of cartography are always changing and often borrow from other technologies. When I began learning cartography at the University of Guelph, Environment Canada, and Michigan State University (MSU) in the 1980s, we drafted in ink on paper and Mylar® sheets and used Leroy® lettering templates. We used Letraset® rub-on symbols when I taught at the University of California at Santa Barbara. We scribed in negative with jewel nibs, opened areas on peelcoat film to define color fills, and applied tiny labels cut from typeset film at MSU and the National Geographic Society. We started to intermix a bit of CAD drawing to automate linework at MSU and then made the jump to Macromedia® FreeHand® software on Macintosh® computers. I began teaching at San Diego State University in 1991 with a full Macintosh lab. The platforms and software have continued to change since I began teaching map design at Pennsylvania State University ten years ago. In 2001, I switched from teaching map design on PCs using Adobe® Illustrator® to using the GIS software ArcGIS® 8.

I built my practical knowledge of the cartographic design capabilities of GIS software through a sequence of experiences that include both mapmaking and teaching. During my sabbatical year at the Census Bureau during 2000 and 2001, I learned how to use many of the design tools available in ArcGIS 8 as I produced an atlas of Census 2000 redistricting data (you will see examples from the atlas referenced in figures in this book). The text and figures for this design book began as material for an online course that I prepared in 2001 and 2002, offered jointly through ESRI® Virtual Campus and Penn State's World Campus, titled *Penn State Editions: Cartographic Design*. I then reworked the introductory cartography course I teach to resident students at Penn State, so they are also learning with GIS tools.

This book describes a subset of the basic knowledge taught in introductory cartography courses. I currently use Terry Slocum's text in my introductory courses and taught for years from Borden Dent's text. Many of us were weaned on Robinson's *Elements of Cartography,* though the fifth and perhaps final edition of that seminal text is dated now. To encourage you to follow this book on design with wider reading, I have included a further readings list of textbooks and other cartographic references at the end of the book. If you would like to follow up on some of my research that I describe, research papers are listed on my Web page at *www. personal.psu.edu/cab38.*

The core tools for good map design are now implemented in GIS software. I encourage you to push the software to make maps more readable and beautiful. I hope this book helps you toward that goal.

Cindy Brewer
State College, Pennsylvania
January 2005

ACKNOWLEDGMENTS

David DiBiase led the collaborative course development effort between Penn State and ESRI and lured me into development of an online map design course in 2001. Suzanne Boden was my first editor for the course material, *Penn State Editions: Cartographic Design,* which became this book. She pushed me past my academic writing style to more straightforward language and separated principles from software-specific information.

Jim Fitzsimmons and Trudy Suchan from the Census Bureau hired me to produce the *Mapping Census 2000* atlas. That project gave me the chance to step back from day-to-day teaching responsibilities and put GIS software to the test for map design and production. Trudy coauthored the atlas and was also a terrific project manager; we got the atlas to press in just six months after the 2000 data was released. Pétra Noble helped us with production, and her questions about type placement, layout, and color helped me understand the sorts of advice GIS analysts want for their own mapmaking.

Linda Pickle at the National Cancer Institute is a pioneer in epidemiological mapping. Our research collaborations and her mapping questions illuminated what statisticians want to know about cartography. Alan MacEachren at Penn State encouraged my work on color and collaborated on research, helping me to see the theoretical underpinnings of my contributions. My ColorBrewer work grew from collaboration with Alan, Linda, and Trudy on an NSF Digital Government grant, and it was assisted by Mark Harrower and Geoff Hatchard.

Judy Olson was my advisor throughout my graduate school days at MSU and is my mentor and friend. She taught me the practice of map design and production and how to make forward-looking use of the ever-changing tools we have. She also taught me how to research mapping and how to change the discipline. Leo and Alfie Zebarth taught me high-end mapmaking during my internship with *National Geographic* magazine. Stuart Allan (Allan Cartography and Raven Maps) taught me about large-format map production back when it was amazingly unwieldy while we produced a map of Santa Barbara with Bill Hunt (Map Link). Bill always keeps me grounded in what's really going on in mapmaking as an industry through his international and wholesaling expertise.

ACKNOWLEDGMENTS, continued

This book is shaped by my students at Michigan State University, University of California at Santa Barbara, San Diego State University, and the Pennsylvania State University. If they get a chance to read this book, they will recognize graphics and advice I first prepared for them. As they ask me questions and I grade their tests and projects, I see how my explanations need to be improved. This book benefits from both their creative ideas and their great mistakes.

The people at ESRI Press have been wonderful to work with. Christian Harder, publisher, welcomed the book into his busy lineup. Edie Punt, my initial editor, improved my writing and is an expert cartographer herself, so she never misunderstood my intent. Judy Hawkins, managing editor, cheered me through the production crunch with fast-paced revisions. Jennifer Galloway, designer, laid out the book in an inviting format and handled all the graphics with skill. It is a treat to create a book on mapping with people who know mapping, graphics, and color, and demand excellent quality throughout.

Charlie Frye at ESRI has been working on improving GIS map design for years. He and I talked about design first at North American Cartographic Information Society (NACIS) meetings, then as we tested ArcMap™ software with the Census 2000 atlas, and recently as a collaborator in both teaching and consulting. He knows the software's capabilities and potential, and his questions are always good ones. I also learn a lot about mapping each time I go to a NACIS meeting. We have a wonderful mix of academics with commercial and government mapmakers. I invite you to join us.

Thank you to David DiBiase, my husband, for both encouraging and weathering this venture. We talk through the future of our field, our big projects, and the little details. He often has a clearer vision of the meaning behind my professional projects than I do, and he is an inspired guide toward the future.

Designing
better MAPS

A Guide for GIS Users

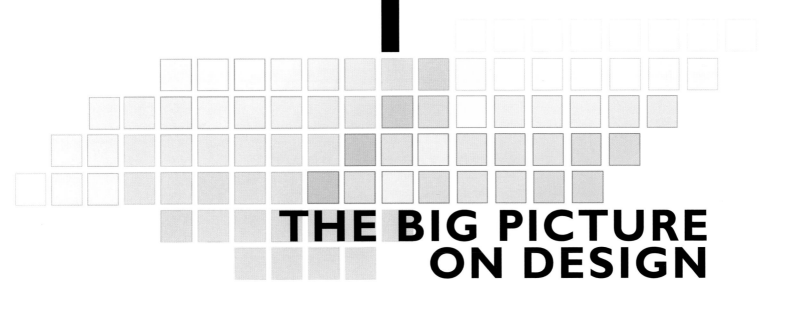

THE BIG PICTURE
ON DESIGN

CHAPTER 1

THE BIG PICTURE ON DESIGN

Every professional knows that communication skills are crucial to their work. Verbal skills are not enough when presenting geographic information—you also need graphic skills. This book helps you develop the graphic skills needed for mapmaking. Cartographic expertise allows you to communicate geographic information clearly with maps. Amateur-looking maps, on the other hand, can undermine your audience's ability to understand important information and weaken the presentation of a professional data investigation.

Designing better maps means thinking carefully about each aspect of the map design process. When creating a page layout, you should size each map element relative to its importance for the map purpose. The positions and sizes of empty spaces between elements are as important to layout as the elements themselves. Designing maps for specific display media and building a polished layout with visual emphasis on key elements of the design will ensure that your audience can read the map and understand your message.

Additionally, being able to transfer map files—to enable other people to build on your hard work or export maps to other software for final production—is the last step in completing a professional design. This chapter presents map design essentials that will help you produce clear, meaningful maps that invite reading.

The essentials of designing better maps are the following:

- designing for map purpose and medium, considering audience, media resolution, viewing distance, and color quality

- linking layout to map purpose, using visual hierarchy, editing decorative design elements, and selecting map projections

- planning a layout, balancing empty spaces, refining alignments in layout, and valuing experimentation and critique

- choosing appropriate export options, including raster formats, vector formats, and maps for the Web

Designing for map purpose and medium

The impetus to design better maps comes from a desire to make maps that are clear and convincing. A successful design begins with knowing why the map is being made. Cartographers begin planning maps by asking themselves and their clients several questions:

- What information is being mapped?

- Who will be reading the map?

- Is the map content coordinated with written text or other graphics?

- What size and medium will be used to display the map?

- What are the time and budget constraints on map production?

The topic and intended audience will dictate many of a map's characteristics. It may be necessary to refer to related research or to other maps in the same field to gauge the amount of detail and relevant symbol conventions suited to the project. Researchers who make their own maps have the advantage of familiarity with their data and how it is typically portrayed. They will still benefit by asking themselves the same set of questions before they begin design work.

Audience

If you are laboring over map design, you are probably making a map for people beyond your immediate work group. Who are these map readers? If the audience is new to the information mapped, they may require a simpler presentation.

Likewise, if they are people who are too busy to spend much time reading, they will also need a simple map that summarizes the information. Maps that have a simple purpose, such as an in-car navigation display showing an address location, demand a simple design. Maps for nonexpert or busy people will have a similar look. They should have a single message that focuses the attention of the reader.

In contrast, maps for people who already know about the topic can be more complex. If they are experts with the data that is mapped, they will expect a rich and multilayered presentation of information that adds to their knowledge or thoroughly supports your (the mapmaker's) contention. The more knowledge and time the map reader brings to the task of reading your map, the more information you will be able to include. More complex maps will motivate advanced map readers to spend more time examining a map on a topic of interest. Detailed information on the map will support their map reading rather than distract from it.

When designing a map, you should also consider your audience's physical ability to read. If the map will be used by older people and others likely to have reduced vision, keep the map text large enough to be legible. If the map will be read in dim or otherwise difficult viewing conditions, use exaggerated lightness contrasts. You may even choose to design your maps to accommodate color-blind readers, who comprise 4 percent of the population.

A map can be tailored to the knowledge level of its audience by reducing the number of categories of data shown. Figure 1.1 shows two municipal water maps made for different purposes. The example on the left shows water mains along with hydrants, meters, fittings, valves, laterals, road centerlines and edges, and sewage mains. This level of detail is suitable for a knowledgeable map reader. The example on the right uses portions of the same dataset, but the map has been simplified to show just the water mains and hydrants. This map would be suitable for a lay audience or a busy city mayor.

Figure 1.1 *Two maps of city water mains. The map on the left includes detail suitable for an expert audience, while the map on the right, with fewer symbols, is appropriate for a novice or busy audience.* Source: New Mexico Bureau of Mines and Geology, Placitas Quadrangle.

The same set of data can be used to make two maps with different purposes by emphasizing different categories of features. The two maps shown below are derived from the Placitas Quadrangle geologic map in New Mexico. The example on the left emphasizes the road to a recreation site, nearby mines, and the network of faults in the area. This map would be suitable as a location map for a group who wanted to plan a field trip to the area to examine these mines. The map on the right would be suitable for a more expert group who is familiar with geologic mapping conventions, the names and ages of geologic formations, and information on strike and dip *(figure 1.2)*.

Maps for different purposes may also have similar levels of detail. The two maps shown below are designed from the same data, but they have different purposes, and so they should have different emphases. They show the same set of lines from a map of Joshua Tree National Park in Southern California symbolized two different ways. In the example on the left, the emphasis is on physical features adjacent to Joshua Tree: the San Andreas Fault (dashed), the transition zone between the Mojave and Colorado Deserts (brown), and sea level (blue). The map on the right emphasizes cultural features adjacent to the park: roads (thin red), the interstate highway (thick red), and populated places (yellow) *(figure 1.3)*.

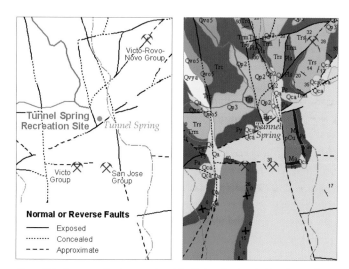

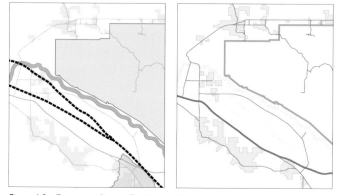

Figure 1.3 *Two maps of Joshua Tree National Park emphasizing physical features (left) and cultural features (right).* Source: National Park Service, www.nps.gov/carto

Figure 1.2 *Two maps showing a portion of the Placitas Quadrangle, New Mexico. The map on the left simply shows faults as well as mining and recreation sites. The map on the right is designed for an expert audience.* Source: New Mexico Bureau of Mines and Geology, Placitas Quadrangle.

Resolution and viewing distance

Choosing how to present a map is part of the design process. Maps are designed for multiple and varied contexts. Each context will be best served by a different map design. Consider a few places we commonly find maps:

- full computer screen viewed at the reader's desk

- computer-projected display presented to hundreds of people at once

- color laser prints distributed to a working group

- black-and-white print for a report that concerned citizens will photocopy at the library

- large plot pinned up at a planning meeting for viewing from across the room

- page in a glossy magazine or book that is professionally printed on an offset color press

- huge backdrop at a trade show

- supporting information in a documentary television show

- black-and-white fax to an emergency response team

- two-inch display on a personal digital assistant (PDA) for route planning

- part of an online interface for Web-based data dissemination

Each of these modes of display places significant constraints on how a map can be made and what it can contain while still being legible. Rather than complain about (or worse, ignore) these constraints, your job as the mapmaker is to use good design to master them. Many of us have attended a talk where the presenter declares that the projector is at fault for the illegibility of the maps. Wrong. The error is made by the presenter who borrowed a design suited for another context or by the map designer who did not account for the final display constraints.

If you need a map in a projected presentation, redesign it with bolder color differences, larger type, and simpler lines to be sure the main messages hold up at coarse screen resolution, bleached by the projector and the room lights, and viewed from a distance. If that map is printed in a book, you can use fine lines, small type, and subtle color differences. If that map will be placed on a Web site and viewed on a computer screen, design the map for screen resolution.

Resolution measures the smallest marks we are able to create within a display. It varies widely among the media on which we display maps. A computer screen may show us 72 dots of light per inch (dpi) across its display. A regular household television has poorer resolution, about 26 dpi for a 27-inch TV (dpi varies with television size; you would need an 8-inch TV for resolution comparable to a computer screen). A laser print may squeeze 600 dots of toner in an inch to build the image. A litho plate on an offset press can reproduce 12,000 dpi from an image-set negative.

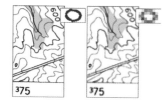

Figure 1.4 *Sample segment of a topographic map shown at 500 dpi (left) and 100 dpi (right). An inset enlargement of a 0 from the 600 contour label is shown at the upper right of each map segment. These enlargements allow each pixel to be counted: the zero is forty-five pixels across 0.09 inches in the higher resolution image and is nine pixels across that same distance at the lower resolution.* Source: Shetlerville Quadrangle, Illinois-Kentucky, USDA Forest Service DRG from *topomaps.usgs.gov/drg/drg_standard_change. html#images*, accessed September 2004.

Map features and type need to be much larger to build them with the emitted spots of light on the computer screen than to reproduce them in ink on a press. A map designed for screen display will look clumsy in a magazine, and a map designed for print may be illegible on screen. There are no bad media, just maps that are not designed appropriately for their media. Your map designs must change to accommodate each medium you are using.

Viewing distance affects map design just as resolution does. Features need to be enlarged to be visible from a distance. Letters two inches high that are seen at a distance of fourteen feet are approximately the same size as 10-point type seen at a reading distance of one foot. A line 2 points wide is practically invisible from across the room, so line widths also need to be increased to retain visibility. (Points are a small unit of measurement used in graphic design; one inch contains 72 points.) Similarly, color differences need to be stark to make small features clear, whether they are small in measured dimensions or small because of the viewing distance.

The following maps show land use in a portion of Clark County, Washington *(figure 1.5)*. A redesigned enlargement of the first map's inset area (the blue rectangle) is shown on the right. The enlarged map uses fine lines and small type that would be suitable for reproduction in print.

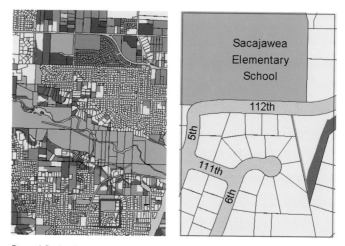

Figure 1.5 *Land-use map of Clark County, Washington. Right: Redesigned enlargement of the inset area (blue outline) from the land-use map on the left.* Source: Clark County land use, Clark County Office, Washington State.

Simply reducing the enlarged inset to its original size demonstrates how the labels become unreadable with coarse resolution. There are not enough dots per inch to represent the small letter forms on this map at screen resolution, and lines have lost their detail and smoothness.

The odd-looking map *(figure 1.7)* shows the reduced version *(figure 1.6)* enlarged back to its original size with no redesign. You can see how poorly the type and lines are represented. You can also see how much information is lost at the coarse resolution.

Figure 1.6 *Poor readability results when the enlarged inset is reduced in size and viewed at screen resolution.* Source: Clark County land use, Clark County Office, Washington State.

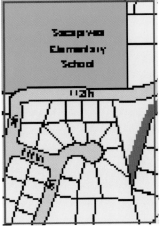

Figure 1.7 *The reduced map has been enlarged to demonstrate the pixelation that makes it unreadable.* Source: Clark County land use, Clark County Office, Washington State.

Figure 1.8 is a redesigned version of the inset map made to display at this smaller size. Both lines and type have been enlarged, improving the legibility of a map this size (compare to the finer lines in figure 1.5).

The larger version is redesigned again with large type and shown at a finer resolution *(figure 1.9)*. This design would be awkward printed in a book to be read at close range, but it would work well for a poster intended to be viewed from across a room.

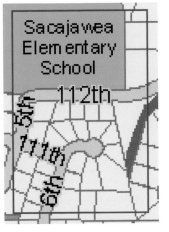

Figure 1.8 *The design has been improved for the small size. The inset is shown at the same scales as figures 1.6 and 1.7.* Source: Clark County land use, Clark County Office, Washington State.

Figure 1.9 *This design is suitable for viewing at a distance.* Source: Clark County land use, Clark County Office, Washington State.

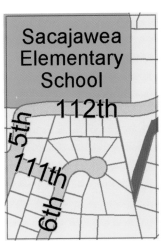

Both resolution of the media and viewing distance determine map design. These examples emphasize how type size and line width must change as a small part of a land-use map is resized and redesigned.

Color quality

Thousands of colors can be produced in print and display environments. Printed pages and cathode ray tube (CRT) screens do a good job of presenting color nuances. Maps designed for display on some liquid crystal display (LCD) screens and projectors require greater color contrast, especially when they include very light colors.

The flexibility of color selection varies widely with media. It is a good practice to test maps in the final media you intend them to be displayed. If the map needs to be readable in widely different media, produce different designs suited to these different contexts. If you want people to be able to make photocopies of a map you are designing in color, test it out on a copier of moderate quality before you finish it. If you want a map to support a presentation, test it with a variety of brightness settings on a projector and look at it from far away with the overhead lights on. Make time to iteratively adjust colors and recheck laser prints of a map before including it in a report to a client. If you are going to spend $50,000 printing a book using professional offset lithographic printing, spend $1,000 on proofs early in the design process to check color sets used in the book. Looking at map colors on a color laser print is not an adequate check of how the offset printed colors will look. You do not want to leave readability of your important maps to chance. Therefore, testing how maps will look in their final form will help prevent many design disappointments.

The maps shown in the next three figures were prepared from the same set of data and base information. They show the change in number of crimes for local police beats in Redlands, California. Each map has a different purpose and thus has different constraints on color use. The first map has six color classes ranging from dark to light to dark, through two hues *(figure 1.10)*. This choice of colors emphasizes the highs and lows and provides details

of change between the extremes. The two hues, blue and orange, represent decrease and increase respectively. The gray roads and white police beat outlines are base information that is readable but does not distract from the main message of change in crime. The readability of the white beat numbers relies on high-quality viewing or reproduction.

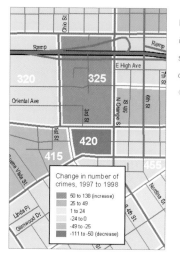

Figure 1.10 *This crime map, which incorporates a detailed data classification scheme using many colors, is suited to high-quality display conditions.* Source: Redlands, California Police Department.

The second version of the crime map is designed for presentation using an LCD projector *(figure 1.11)*. To anticipate differing qualities of projectors and different room lighting conditions, the map has been simplified to emphasize the highest increase and decrease in the area shown. This emphasis on extremes is supported with added text boxes that label the extremes. The beat outlines are also emphasized with a more intense color to be sure they retain readability.

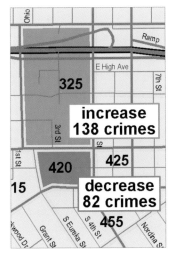

Figure 1.11 *The crime map now has a simpler classification with fewer colors and larger type, which would be appropriate for lower-quality display conditions of a projected presentation.* Source: Redlands, California Police Department.

Figure 1.12 *This grayscale version of the crime map is suitable for black-and-white reproduction.* Source: Redlands, California Police Department.

If the presentation graphics needed to be photocopied in black and white, the dark orange and blue used above would reproduce to grays that were too similar to distinguish. The same map redesigned once more uses only differences in lightness to differentiate between increase and decrease in crime. It is suitable for black-and-white photocopying or laser printing *(figure 1.12)*.

The segment of the Placitas Quadrangle geology map, which you saw earlier as figure 1.2, is shown again *(figure 1.13)*. It has been redesigned here for grayscale presentation.

Design for black-and-white display relies heavily on differences in lightness and variation in pattern. Design constraints for black-and-white media that are unable to reliably produce shades of gray are particularly restrictive. Photocopying and faxing often restrict the mapmaker to black, white, and one or two middle grays for reproduction.

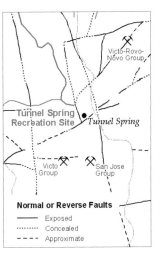

Figure 1.13 *A grayscale version of a segment of the Placitas Quadrangle geology map suitable for black-and-white reproduction.* Source: New Mexico Bureau of Mines and Geology, Placitas Quadrangle.

Linking layout to map purpose

The purpose of your map will determine what parts of it are most important. Which elements of your map do you want people to notice first and remember after they finish reading the map? This ordering of importance—or visual hierarchy—is created by designing some parts of the map to appear as background information and others to take prominence in the foreground. You should design map elements that supply supporting information with decreasing visual importance, echoing their role in understanding the mapped information.

Choosing a map projection is also a design decision that is dependent on the purpose of the map. Projecting the round earth onto the flat page creates unavoidable distortions in the geography of your map. By choosing an appropriate map projection, you can manage the distortion so that it has minimal impact on the message and purpose of your map. The choice of projection partly determines the shape of the map and its layout as well.

Visual hierarchy in layout

A map's purpose determines which of its elements are the most important and should be displayed most prominently in the visual hierarchy. The title and key features on the main map are highest in the visual hierarchy. Supporting information, such as source notes, should be lowest in this hierarchy. Map design is largely a process of deciding how prominent to make each element of your map layout.

Numerous graphic effects can be produced using GIS software. Your decisions whether to use them or change them are guided by the visual hierarchy of information in your map. A clear understanding of the hierarchy of the map's elements to suit its purpose is the essence of good design. Designs that do not follow a logical hierarchy are cluttered, confusing, and hard to read. Map designs that do are crisp, organized, inviting, and to the point.

The list of elements to consider can be extensive for a complex project, though most maps will not include every element:

  ⇐ main map

  ⇐ smaller-scale inset maps showing location

  ⇐ larger-scale inset maps showing detail

  ⇐ insets of locations outside the area of the main map

  ⇐ title

  ⇐ subtitles

  ⇐ legends

  ⇐ scale indicators

  ⇐ orientation indicators

  ⇐ graticule (lines of latitude and longitude)

  ⇐ explanatory text notes

  ⇐ source note

  ⇐ neatline

  ⇐ photos

  ⇐ graphs

Hierarchy is established by an element's position in the map layout, its size, and the amount of open space around it. A note in small text in the lower left corner will be lower in the hierarchy than a title in large text that is centered across the top of the map. Contrasting colors, line weights, and line detail also establish hierarchy.

The elements of a vegetation map of the Democratic Republic of Congo in figure 1.14 are not arranged in any particular order within the layout. This lack of planning produces a cluttered and unclear product. From top to bottom, the elements are the following:

  ⇐ title and location inset map

  ⇐ source note and subtitle for detailed inset map

  ⇐ legend and detailed inset map

  ⇐ scale for main map

  ⇐ orientation indicator (north arrow) and main map with graticule

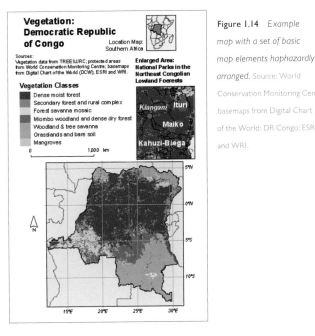

Figure 1.14 *Example map with a set of basic map elements haphazardly arranged.* Source: World Conservation Monitoring Centre; basemaps from Digital Chart of the World; DR Congo; ESRI and WRI.

Below are two organized layouts of the same vegetation map. In addition to the different arrangement of the elements, the visual hierarchy of these layouts is also different. The first map (*figure 1.15*) emphasizes vegetation distributions for the entire country, while the second emphasizes parks located in one forest type. The difference in visual hierarchy between the two maps is established mainly by changing the sizes of elements and repositioning them within the layout.

In the second example, the country vegetation map is an inset rather than the main map. It is smaller and positioned in a less prominent location than in the previous layout. Since this difference changes the apparent purpose of the map, it has been suitably retitled in the map below (*figure 1.16*).

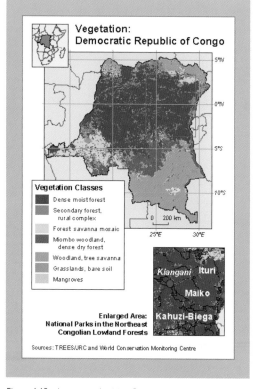

Figure 1.15 *Layout emphasizing Congo vegetation.* Source: World Conservation Monitoring Centre; basemaps from Digital Chart of the World: DR Congo; ESRI and WRI.

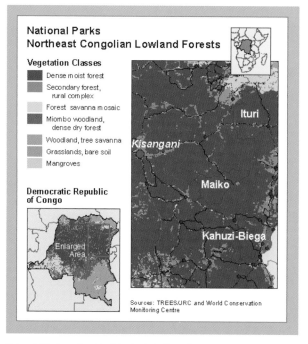

Figure 1.16 *Layout emphasizing parks in Congo forest.* Source: World Conservation Monitoring Centre; basemaps from Digital Chart of the World: DR Congo; ESRI and WRI.

Even small elements can vary widely in their design and their level in the visual hierarchy of a map. A set of scale bars with increasing visual prominence is shown in figure 1.17.

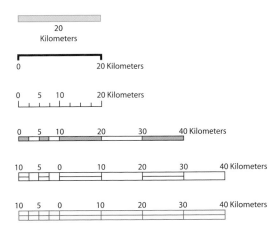

Figure 1.17 *Example scale bar designs.*

The first three scale bars are better suited to thematic maps while the next three are better suited to reference maps. The simple scale bars encourage approximate distance estimates (for example, 10 or 20 km), and they do not distract attention from the map message. The detailed scale bars have sufficient length and segmentation that they let the user calibrate distances and make measurements across the map. A scale bar that is dark, wide, and detailed with many segments (such as the bottom scale bar) is too dominant for a simple thematic map that will not be used for detailed distance measurements. However, the same scale bar may be appropriate for a detailed reference map where its prominence supports a primary use of the map—measuring distances with relative precision between points on the map.

Decorative design elements

Many graphic elements and effects can be used to enhance the primary components of a map layout:

- drop shadows
- line styles for frames
- background patterns
- full compass rose
- zoom lines
- colorful logos
- decorative type fonts
- geometric shapes

An example map of Joshua Tree National Park includes many decorative elements *(figure 1.18)*. They compete for the reader's attention and distract from the main message of the map—ecological zones. They also look fairly silly.

The same map has been redesigned in figure 1.19. The important map content stands out, and supporting information is pushed into the background where it belongs.

Any decorative element can be an effective addition to a map, but only if it is used purposefully. For example, a drop shadow may either attract too much attention away from the map itself, or it may effectively elevate a small, yet important, element to the foreground. Zoom lines that sweep across the page to connect an enlarged area to its location on a smaller-scale map may be vital for one map's purpose, but they may confusingly obscure other

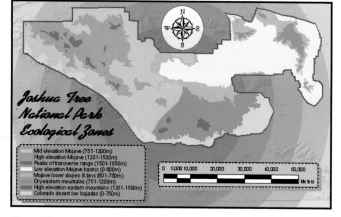

Figure I.18 *Excessively decorated elements on a map of Joshua Tree National Park in Southern California.* Source: National Park Service, www.nps.gov/carto.

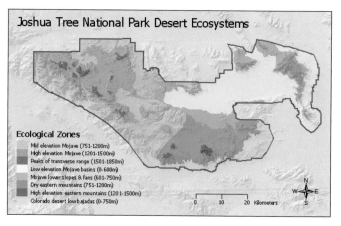

Figure I.19 *An improved visual hierarchy for the Joshua Tree map elements.*
Source: National Park Service, www.nps.gov/carto.

data on a map with a different goal. A simple tiny label, such as "Enlarged area," next to the location on the smaller-scale map may be a better solution if the zoom lines will be too prominent in the map's visual hierarchy.

The enlargement of the northeastern portion of a population density map is elaborately portrayed in figure 1.20. The zoom lines interfere with a portion of the northern megalopolis, which is an important part of understanding population density in the northeast.

The map is redesigned in figure 1.21 with a more subdued cue to the enlargement. Small text explains the relationship between the two areas. The white box and darker blue water link the inset map to its extent, outlined on the main map. Scale bars are included to further emphasize the difference in scale between the two mapped areas.

A background may seem like the one element that should always be lowest in the visual hierarchy, a background in the true sense of the name. But consider a television sportscast—whirling, flashing, colorful, and detailed designs form the background for the information on screen. Should we conclude that colorful and busy backgrounds are the modern way to design information displays? Well, think about how much information is on that display— perhaps four final scores or three performance statistics. Broadcast designers use all that background activity to keep you looking at a screen that has a small amount of information on it. You should be confident that your map contains enough information to attract your reader. Color and detail should be used to make your mapped information stand out, rather than its background. Do not let your background become too high in the visual hierarchy of your map; it is definitely not the most important element of the layout.

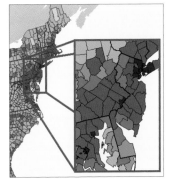

Figure 1.20 *The zoom lines to the inset map showing population in the northeastern United States are distracting.* Source: U.S. Census.

Figure 1.21 *This map uses subtler indicators of the inset map location.* Source: U.S. Census.

Map projections in design

Many mathematicians have been entranced by the interesting puzzle of projecting the spherical globe onto the flat page. My students and I had fun physically acting this out using an old globe that I found on the roadside one day. We stomped and pulled and tore this poor old carcass into a flat surface. We illustrated that all projected maps are distorted in some way (and more kindly). Your job as a mapmaker is to choose a projection that relegates those distortions to places on the map that are not important for your message. This challenge makes projection selection a design decision because it depends on the purpose of the map.

If you are making a detailed map of a small amount of land (a large-scale map), the particulars of map projection will not be crucial unless map readers will be taking detailed measurements from the map. If you are mapping larger areas—all the U.S. states, for example—you should put more thought into the map projection. For continental, oceanic, or hemispheric mapping, projection becomes a critical decision. If you see a map of the United States that looks like a rectangular slab, with a straight-line U.S.–Canada border across the west, be suspicious of the mapmaker's knowledge of map projection and of interpretations of the mapped data.

For example, if you want to understand the road network on a map with a poorly chosen map projection, you will not know whether roads look sparse in an area because it is underdeveloped or because the map is distorted in a way that happens to expand that part of the map. Likewise, maps of point patterns or area densities need equal area basemaps for accurate interpretation.

The map of western Canada below was produced with a plate carrée projection *(figure 1.22)*. The length of one degree of latitude on the page is equal to the length of one degree of longitude, forming a square grid. This projection is sometimes misnamed "no projection." Plate carrée seems like a fine idea until you remember that the length of degrees of longitude get smaller as you near the poles. (The length of one degree of longitude is half the length of a degree of latitude at 60 degrees north.) The provinces and especially the northern islands of Canada appear stretched horizontally because they are distorted by the projection. East-west scale (degrees of longitude) gets larger as you go north on this map. Judging the density of roads or the size of national parks is difficult with a projection that results in such distortions.

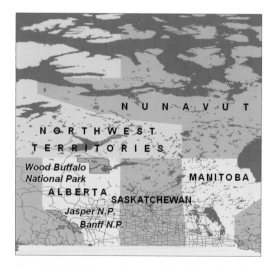

Figure 1.22 *The plate carrée map projection distorts this portion of Canada.*
Source: DMTI Spatial, Inc.

A more suitable projection of the same area of Canada is shown in figure 1.23. This map is made using an Albers Equal Area projection with two standard parallels (lines of true scale) running through the area of interest at 52 degrees and 60 degrees longitude.

The provinces are proportioned quite differently on this map compared to the map in figure 1.22. Wood Buffalo National Park is correctly shown much larger than the parks to the south, as is the openness of the northerly road network. Areas are correct all over this map, so density of features, such as roads and small lakes, can be accurately judged.

Figure 1.23 *Albers equal area map projection of a portion of Canada with standard lines at 52 and 60 degrees north longitude shown as curved gray lines.* Source: DMTI Spatial, Inc.

Projection affects the shape of geographic areas, which in turn constrains the size and layout of the map. You can see how much of northern Canada is not shown in figure 1.23 that was present on the first example. If northern Canada was relevant for this map, a larger frame or a smaller scale would be needed to suit the map purpose.

If you are making thematic maps—special purpose or statistical maps of geographic phenomena such as population density—employ an equal area map projection for most topics. Despite all the fun we could have with projection distortions, this is the important piece of information to remember: If you are mapping data distributions, choose an equal area projection.

If you are mapping the continental United States, the Albers equal area projection, which is customized to the United States, is a common projection choice. You should also make use of the customized Albers projections for Alaska and Hawaii. Each projection has repositioned standard parallels so that no part of the area of interest gets far from these lines where there is no distortion.

Large-scale reference maps often use another category of projection called conformal projections. These projections are better for showing routes and locations because they preserve angles and shapes at points. These advantages come at the expense of preserving areas, and they will misrepresent densities of features in parts of the map distorted by the projection.

Planning a layout

When you are creating a page layout, you should size each map element relative to its importance for the map purpose. Think about the logic of the position of each element relative to other elements. Then step back, squint your eyes, and look at the arrangement of empty spaces on your page. Designing the positions and shapes of those empty spaces is a key to good page layout.

Geographic areas are often irregularly shaped, and a novice designer may be tempted to fill the corners and voids in a display with the remaining elements of the map. Unfortunately, some designs evolve like this—a map designer says, "I see a big hole in a lower corner of my map, so I will use a large compass rose to fill in that problem area." If that sounds like familiar thinking, your future maps will benefit from design practice. The problem with the "fill in" strategy is the resulting overly large or bold map elements that are at the wrong level in the visual hierarchy of the map.

Experimenting with design can reveal new and more effective arrangements of elements in a map layout. Also, knowing how to get the most out of others' critiques lets you finish a project with confidence.

Balancing empty spaces

If the goal of page layout is not consistent filling in, what is it? Page layout is an act of balancing empty spaces. If you have an empty space on the page in one corner, you can position other map elements to produce empty spaces that are similar in size in other parts of the page to balance that gap. These open areas are useful too; they offer a welcome break from the visually dense information of your map and text blocks. They can open up a complex page by separating groups of elements so that their relationships can be better understood.

Two maps of transportation and land use in Prince George's County, Maryland, provide examples of a densely arranged layout and a more loosely arranged layout. Both layouts are suitable for the elements and purpose of the map (*figures 1.24 and 1.25*).

The same maps are marked up to encourage you to focus on the empty spaces in the layouts (*figures 1.26 and 1.27*). In the first layout, the blue highlights are small and similar in size throughout the map. In the second layout, the blue highlights are larger but are still balanced in their arrangement on the page. You can improve your map layouts by learning to see these empty places, create them, move them around, and use them as design elements.

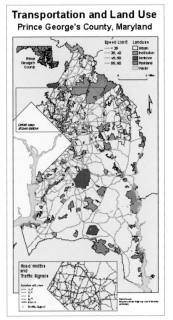

Transportation and Land Use
Prince George's County, Maryland

Figure 1.24 *Compact layout of map elements for Prince George's County, Maryland.* Source: Maryland State Highway Administration, Prince George's County.

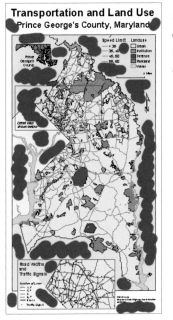

Transportation and Land Use
Prince George's County, Maryland

Figure 1.26 *Blue is used to highlight empty spaces in the compact layout.* Source: Maryland State Highway Administration, Prince George's County.

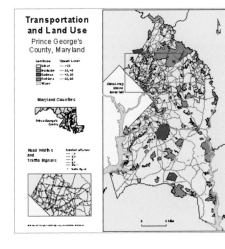

Transportation and Land Use
Prince George's County, Maryland

Figure 1.25 *Looser layout of map elements for the same map of Prince George's County.* Source: Maryland State Highway Administration, Prince George's County.

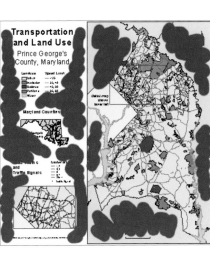

Transportation and Land Use
Prince George's County, Maryland

Figure 1.27 *Blue highlights larger empty spaces in the looser layout.* Source: Maryland State Highway Administration, Prince George's County.

21

Drawing boxes around map elements makes designing with empty space more difficult. A box is not a "bad" design choice for a map, but it can dissect empty space into distinctive shapes—inside the box and outside the box—that become difficult to incorporate into a design. The shape of the empty space outside the box may crowd adjacent parts of the map. Inside the box, the gaps between text and the edge of the box can create shapes that are distracting and difficult to work with. With no box, these shapes coalesce to form a looser space around the text, which has a less distinctive shape and is easier to balance with other empty space in the design.

The top portion of a third layout of the Prince George's County map uses boxes around the inset map, the legend, and the title (*figure 1.28*). Compare this design to the two previous designs.

Purple highlights have been placed in the many crowded and tight spaces created by these boxes in a marked-up version of this same design. These tight spaces can be arranged, but it will be harder to produce a balanced layout (*figure 1.29*).

The purple highlights in the map (*figure 1.30*) emphasize the empty spaces inside and outside the boxes. Notice how little control the designer has over the shapes of these spaces because they are dictated by the boxes.

Figure 1.28 *Boxed elements in the layout of Prince George's county. Notice how the legend and inset have moved higher in the visual hierarchy of the map.* Source: Maryland State Highway Administration, Prince George's County.

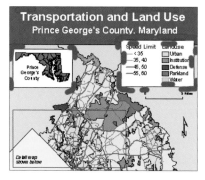

Figure 1.29 *Purple highlights the tight places in this boxy layout.* Source: Maryland State Highway Administration, Prince George's County.

Figure 1.30 *Purple highlights the numerous empty spaces in the boxy layout.* Source: Maryland State Highway Administration, Prince George's County.

In this next example, the boxes create a difficult and distracting set of pinched angles near the center of the detailed inset map from the bottom part of the map layout *(figure 1.31)*.

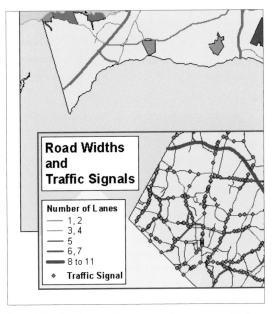

Figure 1.31 *Boxy layout within inset map of transportation details.*

Source: Maryland State Highway Administration, Prince George's County.

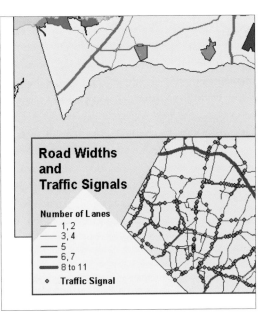

Figure 1.32 *The layout of the inset map is much improved when the boxes are removed.*

Source: Maryland State Highway Administration, Prince George's County.

Empty spaces that flow into each other are much easier to work with when the boxes are removed from the design. Notice also that removing the strong geometric box shapes pushes the legend back in the visual hierarchy where it belongs as supporting information. Strong geometric shapes like rectangles can unintentionally elevate an element in the visual hierarchy of the layout *(figure 1.32)*.

Learn to see and use the empty spaces between elements when you are designing a page layout. Unnecessary boxes around map elements produce gaps and spaces that interfere with the design of an attractive and balanced layout of map elements. It is better to group elements with effective manipulation of empty space rather than by containing them in restrictive and visually dominant boxes.

Refining a layout

A map layout works best when elements that are conceptually related are placed physically near one another. This seems obvious, but in a layout with many map elements, it can be difficult to accomplish. For example, a confusing association can result if a scale bar is placed closer to an inset map than the main map to which it refers.

A layout with many maps, each with explanatory text, is designed well if surrounding empty space unambiguously groups each text block with the map it describes. A general explanation for the entire layout functions well if it stands on its own; not necessarily isolated, but not visually associated with one particular element through proximity.

The location of the scale bar in figure 1.33 is confusing because it is close to three maps, each at a different scale. This layout fails because the importance of proximity is not considered.

An enlargement of a portion of the Congo map, redesigned, shows a scale bar positioned within the main map *(figure 1.34)*. The scale bar is still quite close to the inset map, and therefore might be misconstrued as pertaining to it.

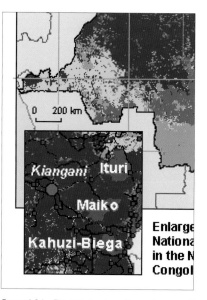

Figure 1.34 *The scale bar is ambiguous because of its close proximity to the inset.*

Source: World Conservation Monitoring Centre; basemaps from Digital Chart of the World (DCW): DR Congo; ESRI and WRI.

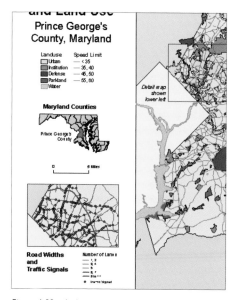

Figure 1.33 *Ambiguous scale bar position between three maps of different scales.*

Source: Maryland State Highway Administration, Prince George's County.

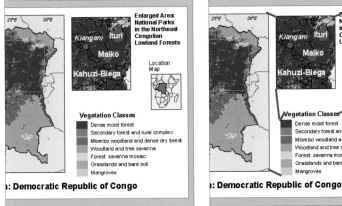

Figure 1.35 *The slight misalignments of elements produces an unpolished layout of this vegetation map.* Source: World Conservation Monitoring Centre; basemaps from Digital Chart of the World (DCW): DR Congo; ESRI and WRI.

Figure 1.36 *The misalignments are highlighted in purple.* Source: World Conservation Monitoring Centre; basemaps from Digital Chart of the World (DCW): DR Congo; ESRI and WRI.

Figure 1.37 *The elements have been aligned to improve the layout of the map.* Source: World Conservation Monitoring Centre; basemaps from Digital Chart of the World (DCW): DR Congo; ESRI and WRI.

As you decide how adjacent objects will be positioned, examine the details of how they align both vertically and horizontally. Look for linear elements that are almost aligned. Do you want them to be perfectly aligned or do they need to be placed intentionally out of alignment? You do not want to unthinkingly align everything; that strategy may produce a display that is more structured and static than is suitable for your map's purpose. Adjusting alignments to be either perfectly aligned or obviously not aligned confirms that your positioning is intentional, not accidental. Ambiguous alignments look like errors.

The portion of the Congo vegetation map *(figure 1.35)* has many elements that are only slightly out of alignment. Though the elements have been positioned reasonably, the map has a messy appearance because these details are not purposeful in their organization. The missed alignments are highlighted in purple in the second map *(figure 1.36)*.

Figure 1.37 pulls the elements into an organized arrangement with intentional alignments. The result is a clear, professional presentation.

Careful alignment can also remedy extraneous or distracting shapes where geographic data coincides awkwardly with graphic frames. These intersections can misleadingly connect the geography to its frame to produce geometric shapes that draw the reader's attention away from the intent of the map. For example, if a state line runs directly to the corner of the frame, the three lines radiating from that one point become visually dominant. Because you can control the position of the geography within its frame, a slight adjustment will usually solve the problem. Choosing different line styles for frames and geographic features can also reduce these effects.

Figure 1.38 shows a few problem intersections between the orange graphic frame and the brown geographic state lines. The problems (highlighted in blue in figure 1.39) include state lines that run along the frame and one that runs directly into the lower left corner. The small tip of Cape Cod that pokes into the inset area at the lower right erroneously resembles an island rather than a peninsula.

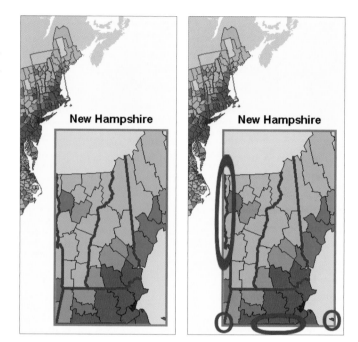

Figure 1.38 *This map contains distracting intersections between the frame and the state lines and coastline.* Source: U.S. Census.

Figure 1.39 *The intersections that can be improved by alignment adjustments are highlighted with blue.* Source: U.S. Census.

Figure 1.40 *Slight adjustments in the relationship between the frame and the geography refine the layout.* Source: U.S. Census.

The geography is not wrong, but the frame positioning is sloppy. Figure 1.40 shows that a slight shift in the frame position relative to the geography solves these problems without compromising the intent of the frame—to show the population density of New Hampshire.

Alignment adjustments are the finishing touches that allow you to create a professional looking page. You can fiddle with them endlessly, so seeing potential problem areas at the start of a project and using guides onscreen and other alignment tools can help you complete your design work efficiently.

Experimentation and critique

In addition to planning hierarchies and balancing empty spaces, a good dose of experimentation often improves a map design. Novice designers tend to place map elements in positions that seem obvious and workable. They may adjust these positions or change the sizes of elements slightly to improve the layout, but they do not question the initial arrangement of elements on the page.

Before you start making small adjustments to improve a layout, push yourself to think of some arrangements radically different from the first one you are assuming will work. Change the page orientation from portrait (tall) to landscape (wide) and see how elements fit together. Move elements from the top of the page to the bottom. Try pulling them into a more compact arrangement with overlapping elements. Overlay titles and text blocks on some conveniently open areas within the map. You may come back to the first layout in the end, but this experimentation is an important first step in map design.

The vegetation map seen in previous examples is shown in portrait and landscape orientations *(figures 1.41 and 1.42)*. Both arrangements are well-balanced with similar visual hierarchies.

Equally important to experimentation is asking other people to judge your draft map layout. When you ask a person to critique your work, your job is to be quiet and let them do what you've asked. A critique is not an opportunity to explain or defend your decisions. You may adjust or discard many of their suggestions, but do that only after you hear them out. During the critique, ask them to elaborate on reasons behind their ideas and interpretations, but do not spend time debating them.

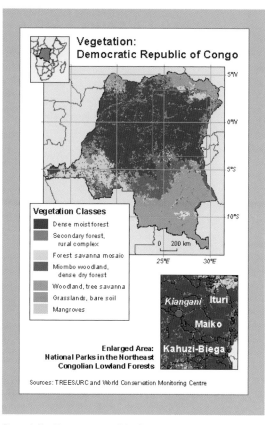

Figure 1.41 *Vegetation map of the Congo in portrait orientation.*
Source: World Conservation Monitoring Centre; basemaps from Digital Chart of the World (DCW), DR Congo, ESRI and WRI.

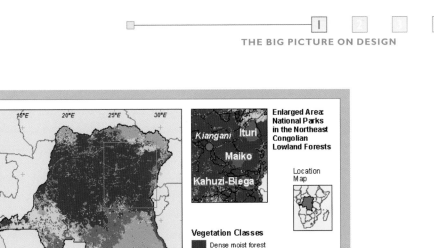

Figure 1.42 *Vegetation map of the Congo in landscape orientation.* Source: World Conservation Monitoring Centre; basemaps from Digital Chart of the World (DCW), DR Congo, ESRI and WRI.

A draft map usually has unfinished aspects, such as incomplete text, nonsense colors, or errors. The person doing the critique will often zero in on these details first. Acknowledge that the work is a draft and encourage them to look at the big picture, the overall layout. Help them get past the details; details are easier to critique than the larger scope of a project.

You should ask a few people for suggestions and balance their critiques. Pay attention to their reasoning and recommendations, but be aware that points of confusion can sometimes be improved by making changes other than the ones suggested. For example, one critic may suggest that legend boxes be made larger so they are more visible and another may suggest spacing the boxes. You may decide that changing the position of the legend so the boxes are not as close to the colorful main map makes them more visible, addressing both concerns without making either suggested change.

A critique is raw material that pushes you to experiment and to refine your decisions. It also keeps you honest—it prevents you from going forward with convoluted adjustments when the overall layout (which at first seemed perfect) has become unwieldy and is ill-suited to the project's current goals.

29

Choosing appropriate export options

Choosing appropriately among many export options lets people without GIS software view and manipulate your map files. This more general concern, which goes beyond the details of mapmaking, allows a wide audience to see the maps you work hard to design.

A map is sometimes only one part of a larger presentation. In order to use a map in a Web or print publication, it must first be exported to a suitable graphic file format. There are a number of graphic formats available, but each falls into one of two categories: raster or vector. Some vector formats may include raster elements as objects within a file.

A raster file uses a regular grid of small cells—called pixels—to store color information across the map surface. It can be thought of as a picture of the original file. The size of cells in this grid determines the resolution: finer grids retain more detail but produce larger file sizes. Individual map elements, including text, are no longer grouped together as digital objects, but rather reduced to collections of pixels. The file can only be altered by editing individual pixels.

A vector file maintains separable objects, and renders their shape, size, and position in the file by connecting locations on the map. Even text characters are built from tiny curves connecting series of x,y locations with mathematical formulae.

The degree to which map objects and text can be edited in a vector file depends on the file type chosen and its associated export options. There are trade-offs between quality, editability, and file size among all export file formats. As with many graphic decisions, testing the suitability of a choice before committing to it is an important step in producing a high-quality final product.

A simple map of Joshua Tree National Park *(figure 1.43)* was exported from ArcMap to seven other file formats. Two files each of the bitmap and JPEG formats were created so that resolution settings could be compared. The resulting list of files is shown here with their sizes *(figure 1.44)*. Notice that the sizes vary dramatically, from 31 KB to 24 MB!

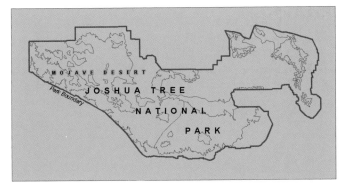

Figure 1.43 *To compare export formats, a very simple map of Joshua Tree National Park (JT1.mxd) with just two line styles, three labels, and a background fill was prepared.* Source: National Park Service, www.nps.gov/carto. Joshua Tree National Park Desert Eco Systems.

Figure 1.44 *Names, types, and sizes for an example set of files exported from the Joshua Tree project (JT1.mxd). Files are in the same order they are discussed in the text of the sections that follow.* Source: National Park Service, www.nps.gov/carto.

Raster export formats

The three most common raster formats for exporting maps are bitmap (.bmp file extension), Tagged Image File Format (TIFF, .tif extension), and Joint Photographic Experts Group format (JPEG, .jpg extension). The bitmap and TIFF formats produce pixel-by-pixel renditions of the map. JPEG is a raster format that is commonly used for Web publishing. It uses a compression algorithm to store a slightly generalized version of the map in a smaller file.

The resolution options vary among formats. To export a bitmap file from ArcMap, height and width of the output file in pixels are specified. When the map of Joshua Tree National Park was exported as a bitmap, the default resolution choice (1,056 × 816 pixels) produced a file with coarse resolution but a relatively small file size (2,525 KB). An enlarged section of the bitmap file is shown in figure 1.45.

Exporting the map again as a bitmap, but this time with three times more resolution (3,168 × 2,448 pixels) produces a higher-quality image *(figure 1.46)*. The type edges are no longer jagged, and the lines are much smoother. The pixels making up this bitmap file are one-ninth the area of pixels in the coarse-resolution version when they are examined at the same map scale. Smaller features can be recorded at higher resolutions. A higher-resolution file will typically come into graphics software with larger dimensions than the coarser version. The image dimensions can be alarming (the file may initially be many feet across when you were hoping to show it on letter-size paper). Raster files can be resized by changing a pixels-per-inch setting without affecting their content.

Figure 1.45 *A portion of the exported Joshua Tree National Park bitmap (JT2.bmp). You can clearly see the pixelation that the bitmap format produces.* Source: National Park Service, www.nps.gov/carto, Joshua Tree National Park Desert Eco Systems.

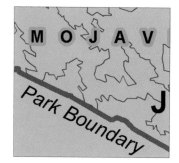

Figure 1.46 *A portion of the Joshua Tree National Park map exported as a high-resolution (3,168 × 2,448 pixels) bitmap (JT2_highres.bmp).* Source: National Park Service, www.nps.gov/carto, Joshua Tree National Park Desert Eco Systems.

But the improved quality comes at a price. The high-resolution setting produces a much larger file—over 22 MB—than the coarse version. To put this size difference in context, you could store more than two hundred coarse-resolution files on a 600 MB CD-ROM, but only twenty-six of the higher-resolution files.

Other choices for file export, beyond numbers of pixels, will also affect file sizes. The color depth setting you choose has an effect on file size, with 24-bit files producing the largest exported files. (Higher depth numbers provide greater numbers of distinct colors in an image because more digits are allotted for storing color information for each pixel in the file.)

To export a map as a TIFF, you specify resolution by choosing the number of dots-per-inch (dpi) in the output file. The default export resolution may be a dpi suited to screen resolutions (such as 72 dpi), which results in a coarse image. The TIFF example shown in figure 1.47 was exported at 296 dpi. It is a high-quality image suitable for publication. Again, this quality came at the price of a large file size (over 24 MB).

All raster file formats share the common characteristic that they are made of only pixels. In raster files, text, lines, and colors are difficult to edit. For example, editing a label would require using graphics software to erase the existing pixels that form the characters and then overlaying new text. Lines and areas do not continue beneath the text and would need to be repaired as well. Changes in the font or style of many map labels would really require going back to the GIS software and re-exporting the map since there are no text objects on the raster map to select and change. To change a map color, every pixel in an area first needs to be selected, working around text and lines that overlay the area. This can be done, but it is much harder than selecting polygons in a vector image and making the color change once.

These raster export formats should be used only for maps that you want to show or print "as-is." This inflexibility can be an advantage when you do not want to pass on a version of your work that can be easily edited or adapted for other purposes.

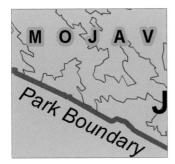

Figure 1.47 *The Joshua Tree National Park map exported as a high-resolution TIFF file (JT3.tif).* Source: National Park Service, www.nps.gov/carto. Joshua Tree National Park Desert Eco Systems.

Vector export formats

The three most common vector formats used to export maps from ArcMap are Enhanced Metafile™ format (.emf extension), Encapsulated PostScript® format (.eps extension), and Adobe Illustrator format (.ai extension). Exported vector files are often much smaller than the raster files discussed in the previous section. Comparing exported vector files of the Joshua Tree National Park map, the EMF file is 42 KB, the EPS is 129 KB, and the AI file is 162 KB. Recall that raster files of comparable quality were larger than 24 MB *(figure 1.48)*. The simplicity of the example map is key. An elaborate file with many small features and numerous labels could readily produce a vector export much larger than a corresponding raster export.

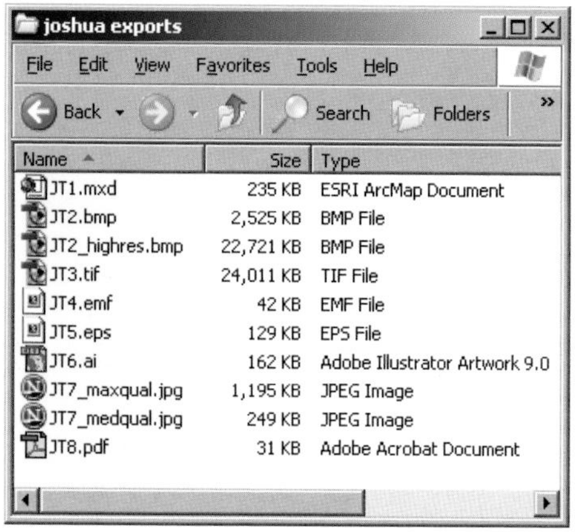

EMF is a multipurpose vector format native to the Microsoft® Windows® operating system. When exported to an EMF file from ArcMap, the example map did not fare well: the type shifted relative to the line work *(figure 1.49)*.

Opening the EMF file in Adobe Illustrator—a vector-based graphics software program—made the map dimensions bigger, requiring it to be rescaled back to its original size. Entire text strings, however, did remain complete and editable. The type was not broken into individual letters or groups of letters, a consequence of exporting to some vector formats. For example, "Park Boundary" can be selected as a single editable object (indicated by the blue line below the letters).

EPS is a common high-quality vector format used to exchange graphics. Exporting to EPS from ArcMap unfortunately causes type strings to break into segments. For example, the "rk" of "Park Boundary" can be selected on its own; the label has been

Figure 1.48 *Note file sizes for the vector formats .emf, .eps, and .ai from the simple map of Joshua Tree National Park.*

Figure 1.49 *Portion of the Joshua Tree map exported as an EMF file (JT4.emf). Letters in the area label MOJAVE are not registered with their halos, and Park Boundary is offset from the line it labels.* Source: National Park Service, *www.nps. gov/carto.* Joshua Tree National Park Desert Eco Systems.

broken into four separate segments. If you needed to edit or restyle this label, the segments would overrun each other or gaps would appear between them. The EPS export may not be a useful option for maps in which type editability is important. As export engines are improved, these types of problems may be repaired, so you may want to periodically do some simple comparison tests like the ones shown in figure 1.48 to investigate format improvements as well as problems.

Exporting to the AI format results in complete text strings and high-quality lines that include Bezier curves between points. For example, "Park Boundary" can be selected as a single object, then edited or restyled. This is important because you want the process of editing text labels to be easy, that is, to not require tedious repairs or manual replacement of individual labels. Map elements are exported as vector objects and can therefore be altered in shape, texture, and color in Adobe Illustrator or other illustration software that can import AI files.

Opening the exported AI file with Adobe Illustrator showed that the character spacing effect used for "MOJAVE DESERT" did not export properly. The letters were no longer registered with their halos. Halos are used to make text more legible over the top of complex line work. They are typically the same color as the background, appearing to "break" the lines before they intersect with the letters. The halos shown in figure 1.51 are somewhat darker than the background, for illustrative purposes. Each halo is exported as an individual vector object, and exporting many halos can create an unmanageably large file. It may be easier to apply halos in the post-GIS editing phase.

Despite the problem with halos and letter spacing, the AI format will be the most trouble-free for publication. You should be aware that some custom type effects and special characters do not export well; therefore, you should always test your choices before relying on them for a design that will need to move beyond GIS software.

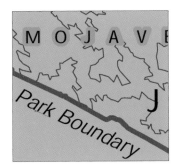

Figure 1.50 *Portion of the Joshua Tree map exported as an EPS file (JT5.eps).* Source: National Park Service, *www.nps. gov/carto.* Joshua Tree National Park Desert Eco Systems.

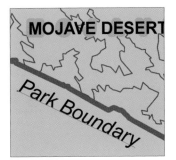

Figure 1.51 *Portion of the Joshua Tree map exported as an AI file (JT6.ai).* Source: National Park Service, *www.nps. gov/carto.* Joshua Tree National Park Desert Eco Systems.

Distributing maps on the Web

There are two export formats well suited for displaying maps on the Web: JPEG in a raster format, and PDF with both vector and raster elements.

JPEG uses a sophisticated compression algorithm to make high-quality raster files smaller. You can see the difference this compression makes by comparing the size of the example (JT7_maxqual is 1,195 KB) to the high-resolution BMP and TIFF files described earlier, which are over twenty times larger. JPEG is a lossy compression algorithm, which means that data is lost when a map is exported as a JPEG. In contrast, a TIFF file has a larger size because all data is retained for the chosen resolution.

When exporting to JPEG from ArcMap, you can control resolution (in dpi) and quality (ranging from low to max). The map shown in figure 1.52 was exported to a JPEG at 300 dpi and maximum quality. The result is a high-quality image with a reasonable file size (1,195 KB). This size is a bit large for Web display, but the quality is good enough that it could be used for some printed contexts.

The second JPEG, shown in figure 1.53, was saved at 300 dpi with medium quality. The savings in file size is good (249 KB) but notice the speckled artifacts around lines and type where compression losses are visible. The reduced file size savings comes at the cost of a poorer-quality image. The file would be fine for many Web applications or Microsoft PowerPoint® presentations, but the quality is not good enough for print publication.

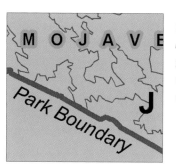

Figure 1.52 *Portion of the Joshua Tree map exported as a high-quality JPEG file (JT7_maxqual.jpg).* Source: National Park Service, www.nps.gov/carto. Joshua Tree National Park Desert Eco Systems.

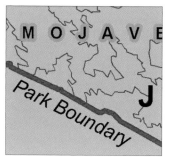

Figure 1.53 *Portion of the Joshua Tree map exported as a medium-quality JPEG file (JT7_medqual.jpg).* Source: National Park Service, www.nps.gov/carto. Joshua Tree National Park Desert Eco Systems.

JPEG files are raster files, so you cannot edit lines and type as objects using illustration software. You should use JPEGs to show your finished work when no further changes will be needed.

Portable Document Format (PDF) is a vector format common for Web display and file transfer. PDF files can be viewed with Acrobat® Reader®, which is available free from Adobe. PDF works well for displaying a high-quality vector image that can be panned and zoomed. The file in figure 1.54 is 31 KB.

You should use PDF for maps with a completely finished design and when you are not concerned about details of type positioning. Notice that the font has changed in the PDF file (compare it to the JPEG shown on the previous page). Single letters, such as the "k" in Park, can be selected. Searchable words are not retained by ArcMap 8 in the PDF because they are exported letter by letter, but ArcMap 9 does a better job with this export format. If you would like to put a map on the Web and have search engines (such as Google®) catalog the title, labels, and descriptive text within the file, you want your map to be exported with complete words and text blocks that can be searched online.

Letter-by-letter type is very difficult to edit, and adjusting font characteristics (such as changing a label to bold) will produce a garbled image. Another problem that may arise with a PDF export is differences in map colors resulting from poor RGB conversions. These problems limit the usefulness of this format as it is currently exported, though improvements in export engines may remedy them. A higher-quality PDF can often be produced by exporting AI files and then saving PDF files from Illustrator or another graphics program.

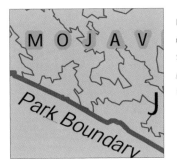

Figure 1.54 *Portion of the Joshua Tree map exported as a PDF file (JT8.pdf).* Source: National Park Service, *www.nps. gov/carto.* Joshua Tree National Park Desert Eco Systems.

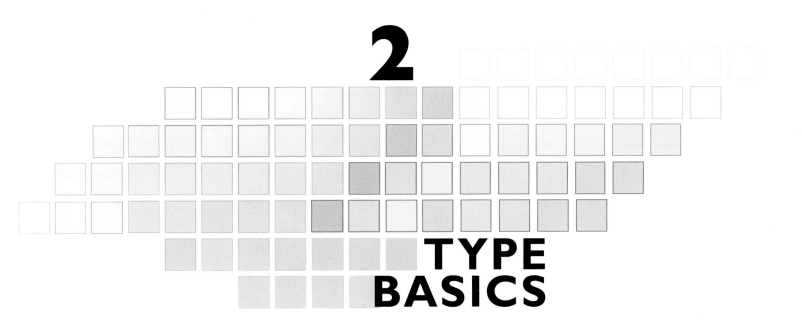

2

TYPE
BASICS

CHAPTER 2

TYPE BASICS

Well-designed type is a major component of the professional appearance and clarity of a map. Cartographers are as careful about how they choose lettering for maps as they are about the symbols they design to present mapped information. Type characteristics both categorize and order features. They establish the personality of a map and its legibility. Type on maps is augmented to improve its contrast and readability against complex map backgrounds. Creating effective maps relies in part on creating effective map type.

Most of the useful type characteristics found in design software can be created using GIS, allowing complete map designs to be achieved in the GIS environment. Fonts and styles, sizes and spacings, and text effects can all be edited for text elements. Mapmakers also need to consider type issues beyond their GIS environment. Type effects export with varied success, and well-planned type can save time in map editing. Understanding font tools in the Windows operating system can assist in using type well.

A mapmaker builds better maps when they understand the following:

□ fonts, including categories of fonts, fonts in the Windows operating system, type styles and font families, and creating special characters

□ label size, including character size, character spacing, and line spacing

□ type effects, including callouts, shadows, and halos

Fonts

Fonts are the personality of a map. They may be serious and authoritative or carefree and inviting. Whatever their tone, they need to be legible in the challenging contexts that are characteristic of mapping. They must be easily read at small sizes and odd angles in varied display media. They are often read among numerous other labels that are close together. In addition, map labels may include unfamiliar words that are indecipherable if they are not clearly legible.

Before you can begin to shape the disposition of your map by choosing a font, you should understand the basic categories of fonts, where they reside in your operating system, and how to access their special characters. You should also be familiar with the potential breadth of font families. This topic describes the basic tools and details you need to know when choosing fonts for mapmaking.

Categories of fonts

Recognizing and naming individual fonts takes much practice and is a more detailed level of study than necessary for most mapmaking. There are, however, three basic categories of font appearance that are useful to learn for map design:

- serif
- sans serif
- display

Figure 2.1 shows a selection of serif fonts with example characters. Serifs are the small finishing strokes on letters. For example, each capital *I* has serifs at the top and bottom of the stroke. The name of each font is shown below the line of example characters. Compare *Qs* to see how the letter forms differ between fonts. Courier is not proportionally spaced like most fonts. Each of its characters occupies the same width along the line, which produces a clumsy appearance in many design contexts.

gqabj GIQWB &?147, Times New Roman

gqabj GIQWB &?147, New Baskerville

gqabj GIQWB &?147, Courier New

Figure 2.1 *Times New Roman®, New Baskerville®, and Courier are examples of serif fonts.*

The next three examples are sans serif fonts, so named because their letter forms lack serifs *(figure 2.2)*. Arial®, Tahoma™, and Futura® are all sans serif fonts, although Tahoma cheats a bit by adding a few serifs to its letterforms, such as the *j,* the *I,* and the *1.*

gqabj GIQWB &?147, Arial

gqabj GIQWB &?147, Tahoma

gqabj GIQWB &?147, Futura

Figure 2.2 *Examples of sans serif fonts.*

You will begin to recognize individual fonts as you learn their general shapes. For example, Futura has very round bowls, slight thinning of strokes (notice how the lines thin as the bowl of the *b* joins the stroke), straight descenders (compare the Arial and Futura *j*), and a small x-height (the lowercase letters are relatively small: compare the sizes of the Tahoma and Futura *q*). With practice, these sorts of characteristics become like the facial features of people you recognize. As you become better acquainted with fonts, you will be able to name the common ones nearly automatically.

Display fonts have distinctive and decorative letterforms whose exaggerated design can be difficult to read. Thus, they have limited use in map design. You may use them in a title or logo, but avoid their use in the body of the map *(figure 2.3)*.

gqabj GIQWB &?147, Tiffany

gqabj GIQWB &?147, Impact

gqabj GI2WB &?147, Kaufmann

Figure 2.3 *Tiffany™, Impact™, and Kaufmann® are examples of display fonts that should be used judiciously.*

A single map will contain few fonts. Usually only two fonts are used, one serif and one sans serif; they should complement each other. Choose two fonts with similar personalities, both informal or both formal in their manner, both modern or both more classic. Fonts may be assertive, refined, official, welcoming, or casual; you will want to choose a pair of fonts that give the right first impression of the map's purpose. Often a serif font is used to label hydrographic and other physical features and a sans serif font to label cultural features. Getting carried away with fonts by including too many or choosing fonts that clash is a quick way to make an amateur-looking map.

Map type is often small, so mapmakers seek fonts with large x-heights, which are more legible. The x-height describes the size of the lowercase letters; it is literally the height of the letter x. For example, Palatino® has a larger x-height than Garamond® (figure 2.4). A consideration when choosing map type is how easy to read small letters are when the map is reproduced and when letters are seen against patterns of tiny dots used to print background colors.

Choosing a font that does not have very thin strokes helps type remain legible after reproduction and against backgrounds. Likewise, thin strokes cause problems with readability at coarse on-screen resolutions. The thin strokes in Garamond are finer than the thin strokes of the same letters in Palatino. For example, the thinning of the lines at the top and bottom of the Garamond o are finer than those of the Palatino o (figure 2.4).

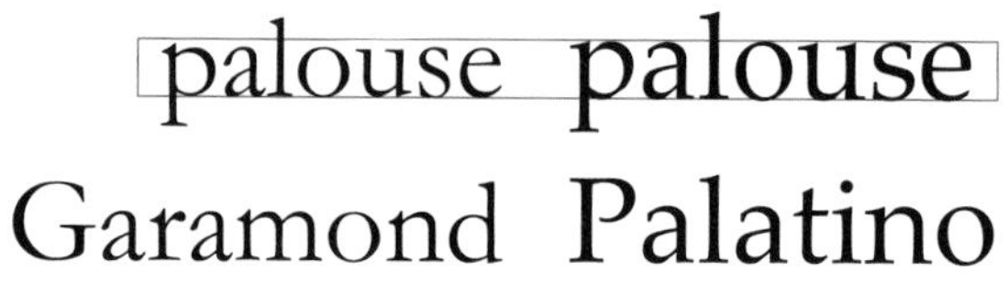

palouse palouse
Garamond Palatino

Figure 2.4   *The purple box around the word "palouse" represents the x-height of the Palatino font, as compared to the smaller x-height of Garamond. Both examples are the same point size.*

The legibility of a selected font is always a concern for mapmakers. Recall the discussions of map purpose and designing for media in the first chapter. If you wish map readers to use your map at coarse on-screen resolutions, select a font that remains legible in that medium. Figure 2.5 shows four fonts with a challenging constraint: small, italic, and angled. Each label includes the word "minimum" above its name as an example with little to separate and distinguish letters. Some of these examples are almost illegible; Goudy®, in particular, does poorly. Others maintain legibility fairly well. Verdana® letters remain more distinguishable at the same point size and angle as the others.

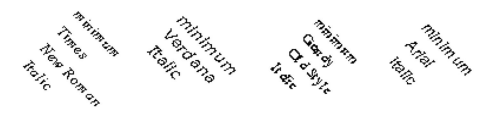

Figure 2.5 *Compare the legibility of the four fonts shown small, italic, and angled. This is a challenging but common situation for map labels. All examples were set at 12 points and captured at 72 pixels-per-inch resolution. A small amount of character spacing was also used to aid readability.*

Fonts in the Windows operating system

You can examine the selection of fonts on your computer using the fonts folder in the Windows operating system control panel. Fonts are installed within the operating system; they are not part of your GIS software. The fonts folder on your computer may include a selection of TrueType® (files with *T* icons, shown below), OpenType® (*O* icons), and PostScript (red *a* icons) fonts. GIS software may be restricted in the formats it can use.

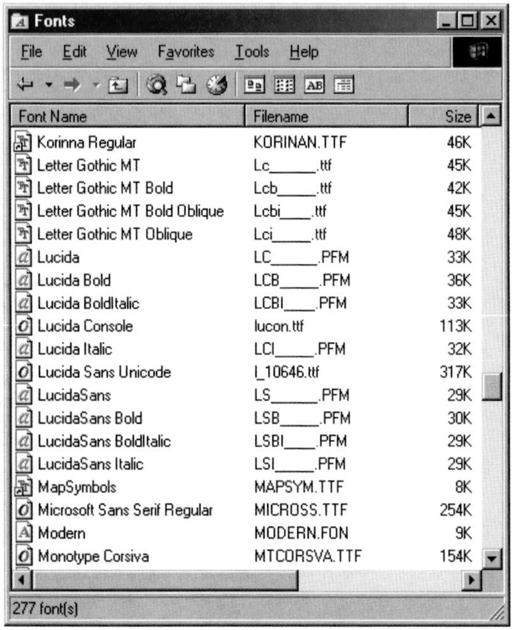

Font file names are fairly cryptic (for example, "Lc" for "Letter Gothic®"). This means searching for font files using file names is rarely productive. You should use the fonts folder, which spells out the font name alongside the file name, to investigate fonts installed on your computer. Font files contain drawing instructions for all of the characters in the font. The computer uses these instructions to draw the characters on screen and to print them. The different font formats, TrueType, PostScript, and OpenType, each require different types of computer processing.

New computers often come with a selection of fonts that is sufficient for standard word-processing and mapping needs. You may never need to add to the fonts installed on your computer. Though it may seem like fonts are free, because many are included in the purchase price of our computers and various software packages, you should know that fonts are intellectual property that are licensed when purchased. To legally produce a map with a customized design, you may need to purchase a font. If you give font files to other people so that they can see your design with the intended fonts, you may be breaking the license agreement you made when you or someone else purchased the font.

Figure 2.6 *Portion of the fonts folder, accessed from the control panel. The fonts folder displays all the fonts installed on the computer, so your fonts folder will contain a different collection of fonts than those shown above.*

One map design strategy is to use widely available fonts such as Arial and Times New Roman. That cautiousness can produce monotonous design work, but it ensures that maps remain readable. If you want type to be a distinctive design element of your map, you may not be able to share the exact digital version of your file if the same fonts are not installed on the computers to which you would like to transfer map files. When you open a map file but do not have all the fonts used on the map, software will often prompt you to substitute another font. Font substitution may disrupt the look and fit of labels and symbols, length and spacing of text lines, and other design aspects that were carefully set within the map based on characteristics of the original fonts. Another option for preserving the look of fonts is to choose an export option that can be set to embed fonts in the file (making the file size larger).

You may be using GIS software that cannot use PostScript fonts, even though they are installed in your fonts folder along with your TrueType fonts. If you want to use a particular font that you only have available as a PostScript font, there are two strategies available to you. You can export the map from the GIS, open it in design software such as Adobe Illustrator or Macromedia Free-Hand, then replace the fonts with the desired PostScript font. Alternatively, you may search for a TrueType or OpenType version of the font to install. Unfortunately, individual fonts may not be available in all formats.

Be critical of the source of your fonts. Fonts are commercial products that vary in their quality of completeness and precision of the drawings of letterforms. Cheap fonts may be poor copies of the original finely drawn characters.

You may be surprised by the way font names vary. Fonts that differ in appearance may have the same name, and fonts that look the same may have different names. These differences occur because different designers and companies seek to differentiate their products, and large companies offer the work of multiple designers and other companies in their libraries.

Figure 2.7 shows three very similar looking fonts. The first two are Lucida®, one offered by Adobe and the other by Elsner and Flake. There is little difference between the two, and it probably would not matter which you purchased. The third is a variation with a similar name and appearance. Lucida Bright looks sufficiently different when the full alphabet is examined; its serifs and thins are different from Lucida's. It would not be an acceptable substitute to match text set in Lucida.

Aa Aa Aa

Figure 2.7 *Three similar-looking fonts: Lucida Serif (Adobe), Lucida (Elsner and Flake), and Lucida Bright (Elsner and Flake).*

Type styles and font families

The regular, italic, bold, and bold italic members of a font family are separate fonts that are installed individually on your computer. Note the four members of each font family in the fonts folder shown below *(figure 2.8)*.

When you select a font from a dropdown list inside word-processing or mapping software, then click buttons for bold and italic, you are actually choosing among four fonts, though only one name for the font displays in the font list. This is an important detail because a person working with a map on another computer will need all four fonts in their fonts folder if the map includes regular, bold, italic, and bold italic styles.

In figure 2.9, you can see that the italic style of Garamond is a separate font whose letter forms are different from regular Garamond. The shapes of the serifs are distinctly different, and the lowercase *a* takes on a whole new form. The italic is a companion font designed to complement the regular version of the font; italicizing is not simply a matter of slanting the characters.

Font families may also include many more weights than simply regular and bold. In figure 2.10, Gill Sans® weights from light through extra bold are shown. These characters represent a set of four separate fonts that are part of the large Gill Sans font family.

Figure 2.11 shows a portion of the extended Arial font family. In addition to these ten TrueType Arial fonts, the larger Postscript Arial family sold by Agfa Monotype includes Light, Light Italic, Light Condensed, Condensed, Bold Condensed, Extra Bold, Extra Bold Italic, Extra Bold Condensed, Rounded Light, Rounded, Rounded Bold, and Rounded Extra Bold.

Figure 2.8   *Portion of a Windows fonts folder showing three font families.*

Garamond Regular
Garamond Italic

Figure 2.9   *Regular and italic forms of the Garamond font.*

Gill Sans, **Gill Sans**, **Gill Sans**, **Gill Sans**

Figure 2.10   *A selection of Gill Sans fonts. From left: light, regular, bold, and extra bold.*

Arial Regular   Arial Narrow Regular   **Arial Black Regular**
Arial Bold   **Arial Narrow Bold**   **Arial Black Bold**
Arial Italic   *Arial Narrow Italic*
Arial Bold Italic   ***Arial Narrow Bold Italic***

Figure 2.11   *Part of the Arial font family (TrueType).*

Special characters

Maps often contain labels with special characters. An obvious example is the degree symbol (°) and the characters representing minutes and seconds of a degree (' and "). These are not the same characters as curly (or smart) single and double quotes (' and "). Some place names have accents on letters or other diacritical marks that you will need to include to produce a correct map.

How do you produce these strange marks and characters? One way is to copy them from the character map in the Windows operating system and paste them into a text element in your map. Below is the Arial character map with the degree sign ready to be copied and pasted into a map.

The character map is also a good choice for examining the many symbols available in specialized fonts that do not include letters and numbers. One of the Wingdings® fonts with icons and bullets is shown in figure 2.13.

Remember to test whether the special characters you use will export to the final format you intend for your map. Worst case scenarios are that the special symbols will not display in the export, or they will crash other software. They should export, but always check before you do too much custom character work.

Figure 2.12 *The Arial character map with the degree symbol selected.*

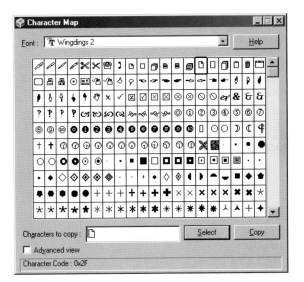

Figure 2.13 *The Wingdings 2 character map with a page icon selected.*

Label size

There are three ways to change the size of a text element, and thereby its meaning, on a map:

- Characters can be set smaller or larger.

- Space between letters can be added so that a text element is wider.

- Lines of text can be spaced so a text element is taller or more compact.

Point size, character spacing, and leading are adjusted to accomplish these changes. This topic describes each of these characteristics.

Character size

Type is measured in points. One point is 0.353 mm or 1/72 of an inch. Paragraphs in a book are typically set in 10- or 12-point type. The miniscule type in an ingredients list on a small food package might be only 3 points, and the large type for a wall map could be 72 points (about one inch high) *(figure 2.14)*.

3 point type

12 point type

72 pt

Figure 2.14 *Relative point sizes: 3, 12, 72 (72-point type is approximately one inch).*

Different fonts can vary markedly in size even when you specify the same point size. In figure 2.15, the x-height of Times® and Batang™ is indicated by the red line. At the same point size, the x-height of Haettenschweiler™ is much larger. In addition, uppercase letters, heights of ascenders (for example, the top part of the letter *b*), and lengths of descenders (the bottom part of *p*) vary in size among fonts. Notice that the Batang font (middle column) has short descenders and wide letters, and that Haettenschweiler has a tall x-height and narrow letters. Setting a particular point size is actually a fairly inexact measure because of the variation among fonts.

Albany Albany **Albany**
Airport Airport **Airport**
Times   Batang   **Haettenschweiler**

Figure 2.15   *Three example fonts set at the same point size.*

You will rarely be able to literally measure part of a 10-point letter and find it to be exactly 10 points high. The reason behind this variation harks back to the origins of type. The "10-point" size refers to the height of the small pieces of metal that carried the raised characters that were inked and pressed to paper. These metal blocks were necessarily larger than any of the individual characters in a font so they could accommodate their full range of shapes. The distance from the top of the ascenders to the bottom of the descenders will approximate the point size. This detailed knowledge is important for two reasons. You will need to experiment if you want to match a font size used in other text related to your map; you will only get an approximate size by measuring the letters. If you prepare a map design with one font and then change to another, it is likely that many of the labels will not fit together as you intended.

If your map fonts will be changed after export (for example, during prepress processing they may be changed to PostScript fonts), be sure to create your design with a font that has an inherent size very similar to the final font that will be used. Do not expect font changes to go smoothly. Selecting fonts carefully before you begin map design work will save you time.

It is often tempting to create very small type on maps so the labels fit closely to very small features. But be considerate of your reader, and remember that some map users have a difficult time reading tiny labels even with their glasses on. If you size type below 6 points, you may be creating an unreadable map. In addition, you need to size type larger if the map will be read from a distance or displayed at coarse resolution.

Character spacing

Character spacing is used to spread a label across an area on a map. Just as it sounds, character spacing increases the distance between letters. The example below shows a label with wide character spacing meant to span an urban area on a map.

Figure 2.16 *Text with 200-percent character spacing.*

GIS and design software allow you to set character spacing (also called tracking or kerning) either in absolute points or as a percentage of the label point size.

It is common to increase character spacing slightly for map labels to improve their legibility at small sizes. A small increase in spacing can also improve the appearance of curved type by making letters less likely to tilt into one another, as shown in figure 2.17.

Figure 2.17 *Adjust spacing slightly so that letters are not crowded on curves.*

In ArcMap, the units for character spacing are expressed as a percentage of the text element's point size. In the DEARBORN HEIGHTS example *(figure 2.16),* the character spacing of the text is set to 200 percent, which means approximately two letters will fit in the spaces between letters. Given the variability of character size for different fonts, this is an approximate measure that requires some experimentation to get the desired effect.

Line spacing

Leading (pronounced "led-ing") is the spacing between lines of type. Leading may be specified as the distance between baselines, as a percentage of point size, or as an absolute amount by which spacing is increased. Large leading values are used to spread out a stacked areal label to express the extent of the region identified. Leading is often combined with character spacing for this purpose.

It is common in map design to use slightly less than the default leading to move lines closer together. For example, when multi-line labels appear in crowded areas of the map, tight leading can unambiguously link them to their point features (*figure 2.18*).

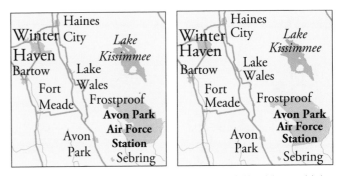

Figure 2.18 *This map pair shows labels with regular leading (left) and the same labels with tighter leading (right).* Source: U.S. National Atlas using U.S. National Atlas data.

When you set a block of text with multiple lines, the standard leading is about 120 percent of the type size. For example, lines of 10-point type measure about 12 points from baseline to baseline, leaving about 2 points spacing between the bottom of the descenders and the tops of the ascenders. In ArcMap, the default leading setting of zero points approximates this standard 120-percent line spacing (*figure 2.19*).

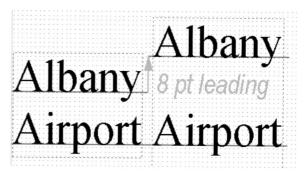

Figure 2.19 *In ArcMap, 10-point type (enlarged) with 0-point leading creates a 2-point gap between letters (dot grid has 1-point spacing).*

Figure 2.20 *The leading for the Albany Airport label was increased by 8 points. If you count using the 1-point grid, you can see that there is now a 10-point gap between the bottom of the descender (the y in Albany) and the top of the uppercase letter (the A in Airport).*

Although leading is set in points in ArcMap, the point size does not reflect the baseline-to-baseline measurement. Instead, it indicates increasing or decreasing spacing from the default. For example, if you set leading at 8 points, line spacing increases by 8 points. The two blocks of 10-point text in figure 2.20 are enlarged, and the leading increase of 8 points is indicated in orange. The result is approximately 10 points of space between the two lines.

Figure 2.21 compares the result of leadings set at -2 (decreasing line spacing), 0 (no change from default), and 8 (increase).

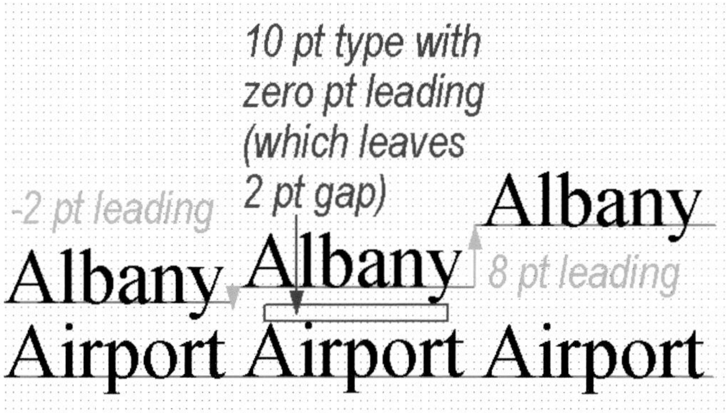

Leading is an approximate measure for the same reasons that different fonts have different inherent sizes, even when set at the same point size. Figure 2.22 shows the same leading setting for the Palatino, Times, and Haettenschweiler fonts. The baseline-to-baseline distance for Times is indicated by a green box. Notice that the leading for Palatino is larger and the Haettenschweiler leading is tighter even though each is set at "zero leading."

Figure 2.21 *Enlarged view of 10-point type with three leading settings: -2, default (0), and 8.*

Figure 2.22 *Note the differences in line spacing for these three fonts. All three are set to the same point size with zero leading in ArcMap 8.*

Type effects

In addition to the basic type characteristics of font, size, style, and spacing, there are a variety of effects you can use to augment text on maps. Callouts clarify the link between a location and its label, which is especially useful in densely labeled areas. Shadows and halos improve the contrast between text and nearby map features. Callouts, shadows, and halos do not change the form of characters in a text element. They are additions to the characters. Thoughtful use of type effects improves map clarity and enhances design.

Callouts

Callouts use graphic elements such as leader lines that explicitly link a text label to a point location. Use callouts sparingly; reserve them for when you need to identify points in a densely labeled or otherwise inaccessible location on the map. Putting a callout around every label on a map produces clutter in most contexts.

A leader line can connect to any part of the label: the beginning, the end, or the middle—whichever is closest to the feature. The goal is to use leader lines sparingly and to make them as short as possible *(figure 2.23)*.

A callout styled as a cartoon balloon is too dominant for most mapping contexts, though in journalistic mapping it can be appropriate to highlight a few locations discussed in a story. But a filled callout blocks underlying map information. GIS and graphics software provide a lot of flexibility in changing the design of callouts *(figure 2.24)*.

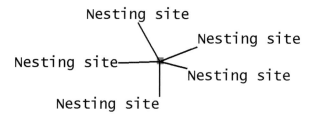

Figure 2.23 *Leader line positions were dynamically selected by ArcMap as the "Nesting site" label was moved manually around the feature. (Only one of these labels would appear on the final map.)*

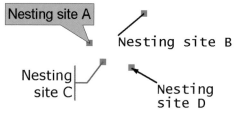

Figure 2.24 *The "Nesting site A" label shows ArcMap's default callout style with color adjustment. Three other styles are suggested.*

Shadows

Shadow effects, when skillfully applied, increase the legibility of text on maps. A shadow is simply a graphic copy of the shape of the characters, offset, and rendered in a contrasting color to enhance text prominence. It is defined by the offsets applied, which adjust the perceived height of the text above the page (*figure 2.25*).

REVISED **REVISED**

ZONING **ZONING**

Figure 2.25 *Emphasis has been added to text using a gray shadow with a significant 2-point offset to the right and down. The result is that the text appears to be floating.*

Figure 2.26 *This thin shadow will provide subtle contrast against other map detail. The result appears to be a solid three-dimensional block. On a crowded map, this style of shadowed text is more legible than the floating effect of the previous example.*

A smaller shadow can increase legibility of text by adding contrast with background colors and features. The shadows shown in figures 2.26 and 2.27 are offset only 0.5 points right and down. These thinner shadows enhance the text but are less distracting than shadows that make the text appear to float above the page.

Figure 2.28 demonstrates how shadows improve contrast between both light or dark type and a multicolored background of vegetation symbols. Without a shadow, the white label is hard to read over light areal symbols, and the dark label is hard to read over dark areas of the map.

Use shadows sparingly in map design, and remember to check how a shadow looks in the final medium for the map. You will also want to be sure that a shadow effect exports properly if your map will be moved out of the GIS environment. Vector export formats may omit, offset, or reshape the shadow letter forms.

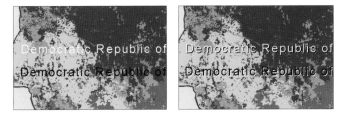

Figure 2.28 *Shadows below white and black text improve contrast against a multicolored vegetation map.* Source: World Conservation Monitoring Centre; basemaps from Digital Chart of the World (DCW), DR Congo, ESRI and WRI.

Figure 2.27 *This thin, white shadow, made with 0.5-point offsets, increases the contrast between the text and a multicolored background.*

Halos

Halos function much like shadows. A halo can be used to decorate text or to subtly improve contrast with the background. The halo effect is also called "letter casing" by cartographers and an "inline" in some design software. In the bottom example of figure 2.29, a halo fill color has been used that matches the background color. The halos break lines where they are close to text so that text remains legible.

Figure 2.30   *A 0.5-point halo around 6-point type (enlarged) improves legibility against a cross-hatched background.*

Figure 2.29   *Two-point halos on 10-point type (enlarged) break the yellow lines and increase legibility.*

The halo in the figure above is bold to demonstrate the effect. Smaller text and subtler breaks call for a thinner halo. For example, a 0.5-point halo on 6-point type is a useful combination. Notice that this size halo fills most open areas within letters and small spaces between letters, masking most of the cross-hatching where it might interfere with character legibility but without masking too much of the area around the text *(figure 2.30)*.

The goal when selecting a halo size is to clean up small pieces of line or other content that show between letters while masking as little of the underlying map information as possible.

In figure 2.31, the legibility of street names set in 4-point Verdana is improved by offsetting the labels 0.5 points away from the road feature. A thin 0.5-point halo was then applied to the labels to mask interfering road segments. The challenge is to choose a halo size that is an adequate compromise between improving legibility while not obscuring too much data. For example, the halo here is not quite large enough to mask the road segment that appears between the *r* and *f* of the vertical Butterfield St label in the top-center of the map.

TYPE BASICS

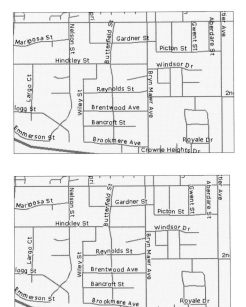

Figure 2.31 *The top map has no halos. Notice the interference of street lines with the label letters. Thin halos that produce breaks in the street lines for the labels were used on the bottom map.* Source: Redlands. California Police Department.

Figure 2.32 *Text that borrows its color from underlying features is unreadable on a vegetation map of the Congo.* Source: World Conservation Monitoring Centre; basemaps from Digital Chart of the World (DCW): DR Congo; ESRI and WRI.

Figure 2.33 *Halos that contrast with both the text and the background improve legibility.* Source: World Conservation Monitoring Centre; basemaps from Digital Chart of the World (DCW): DR Congo; ESRI and WRI.

On the other hand, it seems some of the halos should be a bit thinner because lines are being unnecessarily nibbled near the bottom of letters. For example, "Gwent St" and "Aberdare St" labels (top right on the bottom map) chew into the streets they name. This problem could be improved by manual repositioning of the labels or by adjusting the automatic placement specifications for labels. The contradiction of halos that seem both too large and too small suggests that a suitable halo thickness has almost been achieved. If you were preparing this map for print, it would be important to check halo widths in a high-resolution print because small sizes are rarely rendered accurately on screen.

Map labels sometimes need to span many background colors, both dark and light. This varied contrast makes it difficult to select a single color that remains readable. As with shadows, halos that contrast with the text and background colors (rather than matching the background) can improve legibility. For example, type that is the same color as one of the vegetation symbols is nearly invisible. Halos add the contrast necessary to read the labels easily *(figures 2.32 and 2.33)*.

Halos are useful for mapping because they can help ensure that text is legible on varied backgrounds. They also have drawbacks. They do not export well, they increase processing times, and they can produce excess visual noise around labels. Trying to create subtle halos that match the background can be a challenge if multiple background colors are used on a map. Use of contrasting fill, line, and type colors are alternatives to haloing type, so do not become overly reliant on this useful effect. ArcMap's advanced symbol-level masking is also an alternative, particularly when labels overlay multiple line and polygon symbols.

59

3

EFFECTIVE TYPE IN MAP DESIGN

CHAPTER 3

EFFECTIVE TYPE IN MAP DESIGN

Labels are an inherent part of a map's symbology. The technical challenge of making labels is complemented by the analytical role they play on a map. Thus, clear labeling helps your audience correctly interpret the map data.

Clear labeling follows conventions of placement that vary for point, line, and area features. Although it is faster to set type characteristics for groups of features all at once, there will be times when you need to edit individual labels to improve the clarity of the map. Cartographic conventions for placing labels often conflict when numerous labels appear on a map dense with features. Making sensible trade-offs between labeling conventions is often a matter of knowing how to follow a convention and knowing how to bend it. Knowing the rules lets you break them less often and more skillfully.

A competent map designer makes better choices for label appearance and placement by understanding the following:

- map text, including differences between graphic map text, dynamic labeling, and annotation

- labels as symbols, indicating feature location, category, and hierarchy, and using ambiguity and contradiction in classification with type

- label placement conventions for point, line, and area features, as well as dense label placement and trade-offs between placement rules

Map text

Your decisions about handling text should be influenced by how much text will appear and the role that text plays in the map. You may be positioning individual text elements in static positions, generating labels for thousands of features at once, or creating feature-linked annotation. Your goal is to understand these three kinds of labeling and be able to use them together for efficient and skillful map design.

The descriptions in this section make use of some ArcMap terms (such as data frame, attribute table, and annotation), but the principles yielding these three kinds of labels—graphic, dynamic, and annotation—are generally applicable in GIS use. In contrast, map labels created in illustration software behave mostly as graphic text.

Graphic map text

When you want to add general information to a map, such as titles, subtitles, sources, or notes, you place them as graphic text elements in the layout. These text elements are not associated with particular map features or data frames in ArcMap. They remain the same size and in the same position as where you originally placed them, unless you drag them to a new location in the layout *(figure 3.1)*.

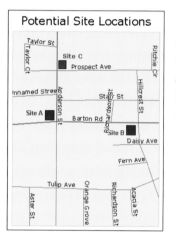

Figure 3.1 *The title, placed as graphic text, will stay in a fixed position in the layout regardless of changes to the geographic extent shown in the data frame.* Source: Redlands, California Police Department.

Technically, you can add graphic text to a layout to label map features, but this can be problematic. The labels will not be associated with the data frame, so they will not move when the map scale or the extent shown in the data frame changes. Thus, the label may no longer be near its feature as the map is edited. Adding graphic feature labels to a layout is fine if the scale and extent of the underlying data frame is fixed and it contains only a few labels. Custom text editing is straightforward for these kinds of labels.

Figure 3.2 shows the disadvantage of adding feature labels as graphic elements to the layout rather than into the data frame. Panning within the data frame (the yellow area) reveals more of the streets to the north of the three mapped sites. Notice that the labels and symbols for two sites (sites A and B, in blue) stayed linked to their geographic locations. The text for these sites was added into the data frame. The Site C label is a graphic element that was added to the layout. It remained in the same place on the page, not moving with its geographic location at the corner of Prospect and Anderson.

Figure 3.2 *The Site A and Site B labels and symbols moved when the data frame was panned. The graphic text element for Site C remained static on the layout page because it was not added to the data frame (compare this map to figure 3.1).* Source: Redlands, California Police Department.

Dynamic labeling

Dynamic labeling allows you to automatically label multiple features in a layer, such as all roads in a road dataset that is represented on a map. If your GIS attribute table does not include a field populated with road names, consider adding a new field to the table before you start the process of placing individual feature names. It may take you longer to create labels as individual graphic text elements for each road than it would to enter them into a new field.

When you use ArcMap to create dynamic labeling, there are numerous settings that allow you to define classes of labels, set weights and buffers, allow overlapping labels, control duplication, constrain placement conventions, set the scale at which labels are shown, and set type characteristics. For example, to label all streets using 10-point Verdana with 1-point halos, set those parameters in the dynamic label properties for the streets layer. This is much faster than selecting and setting the characteristics of each street label individually. The result will be a reasonably complete set of map labels that are automatically placed on the map. As you change the map scale and extent, the positions of map labels will dynamically change. At smaller scales, when there are many features to display, the software will label fewer features in a given area than at larger scales. In a dynamic map-use environment, not every feature needs to be labeled because the user can interactively query them using GIS tools.

If fixed label positions are important for a particular project (as they often are), you can customize the position and characteristics of text beyond the settings available for dynamic labeling using annotation in ArcMap instead.

Figure 3.3 *When you pan across a map with dynamic labels, you see why they are called "dynamic." The labeling software repositions, adds, and removes labels to fit the available space. Compare the first map to the second one: Starr St, Barton Rd, and Daisy Ave labels have moved.* Source: Redlands, California Police Department.

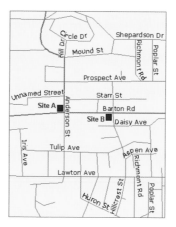

Figure 3.4 *At a smaller map scale the software dynamically selects fewer representative labels. Many streets do not have labels in the view but may be queried by interacting with the GIS software.* Source: Redlands, California Police Department.

Annotation

Unlike dynamic labels, the position of annotation is fixed relative to the data, and it can be manually repositioned to a preferred location. Annotation can be edited, either manually or, in the case of feature-linked annotation, by updating the corresponding field in the attribute table. Feature-linked annotation will move if its feature is relocated. Annotation cannot be moved outside of the data frame, so it is not a suitable format for titles and notes that belong in the area surrounding the map.

Labeling maps is one of the most time consuming aspects of cartography, so you will probably want to use all three kinds of labels discussed here. To efficiently produce a well-designed map in ArcMap, first use dynamic labeling to create a complete set of labels in generally correct locations. After setting as many global characteristics as you can on entire categories of labels, convert them to annotation. Annotation can be edited, so you can now refine placement and type characteristics for individual labels. Add graphic text only for marginal elements. This sequence lets you save time and produce a well-designed map by editing as few labels as possible.

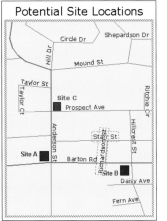

Figure 3.5 *The dynamic street labels were converted to annotation. The three selected pieces of annotation are being repositioned: Shepardson Dr has been moved left, and the overlapping Richardson St and Starr St labels will be moved apart.* Source: Redlands, California Police Department.

Figure 3.6 *The positions of Shepardson Dr, Richardson St, and Starr St have been adjusted. The text has then been customized for the map topic. Street annotation near the three potential sites has been emphasized with bold type.* Source: Redlands, California Police Department.

Labels as symbols

By categorizing feature labels, you help the reader find features on the map and draw conclusions from them. Sensible groups containing fewer features are easier to look through than large collections of uniformly labeled names (*figure 3.7*). Characteristics of the type help map readers identify each kind of feature. Groups of features may be distinguished by differences in symbols, but varying the type used for feature labels emphasizes these differences.

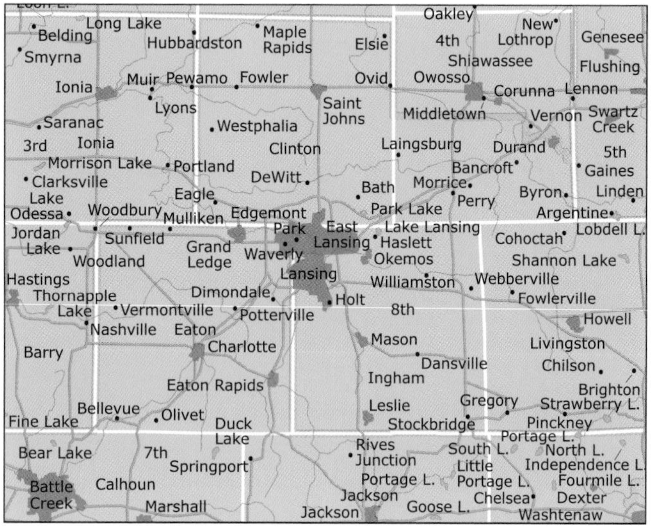

Figure 3.7 *Lack of type differences between categories of labels make this map hard to read. Numbers associated with congressional districts (pink boundaries) are difficult to find and interpret. Differences between lake and town labels are also difficult to sort out, especially when a town has a name like Lake Odessa (middle left). See figure 3.13 for an improved version of this map.* Source: U.S. National Atlas, www.nationalatlas.gov.

Indicators of location

While we usually think the function of labels is to name locations, they also help your readers find locations. For point features especially, the labels are generally larger than the symbols, making them easier to find than the symbols. This is why mapmakers follow conventions for label placement. The relationship between a location and its label should be as predictable as possible, so the map reader does not need to work hard to determine which label belongs to which feature.

How you align a multiline label with a feature location is guided by the need to use the label as a location indicator.

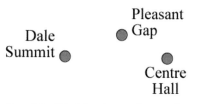

Figure 3.8 *Right, left, and center alignments for two-line labels positioned appropriately next to each symbol.*

The Dale Summit label is aligned right-justified so both lines of the label crowd near the point being labeled. The Pleasant Gap label is aligned left to indicate the point is to the left. The two lines of the Centre Hall label are centered underneath the feature point. Alignment gives clues to where the point can be found relative to the label.

Setting a consistent text alignment based on the label position facilitates map editing. You should right-align all labels positioned to the left of their point, even if they have only one line and you have already positioned them next to their point. Likewise, left-align all labels positioned to the right and set all centered labels to centered alignment. This keeps the proximity between the point and the label constant, so if you change the font or size of the labels, you will not need to reposition each one. The same follows for line and area feature labels. Your goal should be to build a robust map that will withstand changes in type without falling into disarray.

Labels are shown for a set of five point locations *(figure 3.9)*. Each has an anchor at the lower left corner of the text box (green dot), indicating that they are all currently left-aligned.

Figure 3.9 *A selection of left-aligned labels that are seemingly well positioned next to point features. Green dots mark the anchor for each text box.*

When the labels are enlarged, the bottom left corner of each text box remains stationary relative to the data. The larger characters extend from that point, overlapping their symbols and even other labels *(figure 3.10)*.

Figure 3.10 *Resizing carelessly aligned labels causes some labels to overlap their symbols.*

The original set of labels is shown again below with the anchor points (identified by blue dots) placed at the corner closest to the symbol *(figure 3.11)*.

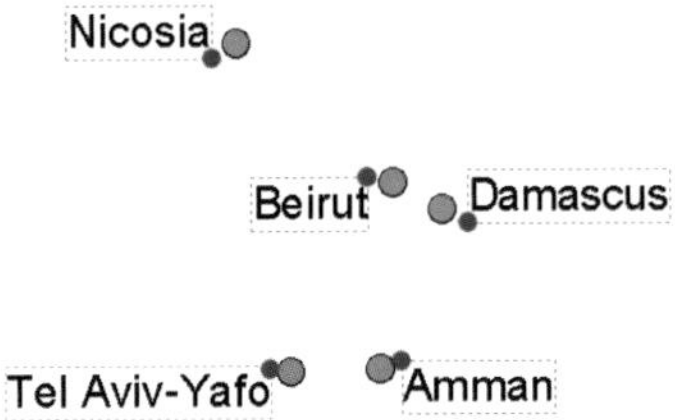

Figure 3.11 *Labels are carefully aligned both to left or right and vertically (anchor points are shown with blue dots).*

Improved anchor point positions result in improved label positions when the labels are again enlarged *(figure 3.12)*. Placements could be refined, but doing so would take much less time to perform than the edits required in the previous example.

Figure 3.12 *Carefully aligned labels can be resized with much better success, potentially saving hours of editing.*

Indicators of feature category

Variations in the type style of feature labels can be used to categorize the features themselves. This approach organizes the components of your map and makes it easy to understand. A reader will be able to find specific features quickly if type styles thoughtfully categorize feature types. For example, consider a map with 1,220 labels. If only twenty of those labels are river names, it will be easy for the reader to find a specific river if you can direct their attention to just those twenty labels using a distinctive font, style, or color to differentiate rivers from other features (figure 3.13).

The challenge is to not inadvertently impose a hierarchy on your labels when you categorize them by type style. Characteristics that you can use to categorize features, without suggesting a difference in their magnitude or importance, are the following:

- font
- posture—roman (regular) versus italic
- color hue
- arrangement

The two labels in figure 3.14 use different fonts. The landform name is a serif font and the school name (Penn State) is sans serif.

Nittany Mountain
Penn State

Figure 3.14 *Different fonts are often used to indicate different categories of features.*

A professional looking map generally relies on no more than two font families. So, you should use fonts to distinguish only very broad categories. It is common to use a sans serif font to label all cultural features (towns and landmarks) and a serif font to label all physical features (mountains and hydrography). With only a few font families on the map, you must rely on other type characteristics to establish additional feature categories.

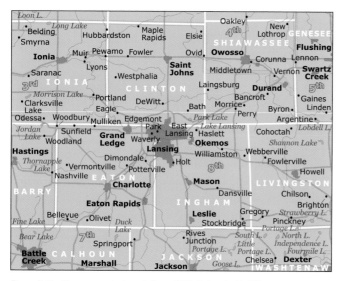

Figure 3.13 *The same map shown in figure 3.7 is designed with different type characteristics for categories of features. For example, different hues are used for congressional district numbers (pink) and lake labels (blue). An italic serif font is also used for the lake labels, which makes them easier to search through because they are visually separate from the town labels.* Source: U.S. National Atlas, www.nationalatlas.gov.

Roman (regular) and italic versions of the same font are shown below. The creek label is italic and the street label is roman. Italic postures are harder to read than roman. They are designed to add emphasis to text by slowing the reader down. Use them to label categories with fewer or less important features *(figure 3.15)*.

S. Water Street

Spring Creek

Figure 3.15   *Posture (regular and italic) indicates different categories of features.*

Label color can further accentuate category differences already established by a type style difference *(figure 3.16)*.

S. Water Street

Spring Creek

Figure 3.16   *Augmenting a style difference by hue further differentiates between feature categories.*

The arrangement of characters can also categorize features. Characters arranged to follow a line indicate a linear feature. The difference in arrangement between horizontal and curved labels in figure 3.17 implies two different kinds of features, even though the features are not shown. The curve not only indicates location, but also categorizes its feature.

Variation in hue is used in figure 3.18 to further distinguish two categories already established by the arrangement of the characters. Type characteristics can be combined either to emphasize existing categories or to subset categories into smaller divisions.

Figure 3.17   *Labels that follow a curve indicate a different category of feature than horizontal labels.*

Figure 3.18   *Augmenting arrangement with hue to emphasize feature categorization.*

Case, leading, and character spacing may also be used to indicate categories of features. For example, uppercase letters may indicate an area feature, and lowercase letters may be used for a point feature. Spacing between lines and characters can allude to an area feature. These three characteristics do not provide as clear a categorization of features, though, because they can also suggest differences in hierarchy.

Indicators of feature hierarchy

Within one category of feature there are often differences in magnitude that you would like to highlight. For example, cities have varied populations and rivers have different volumes of flow. Use type design to describe these hierarchical relationships.

You learned about hierarchy in the first chapter as it applied to map layout. When the notion is applied to type, it is also linked to map purpose. The more important, often larger, features should be more visually prominent, or higher in the visual hierarchy. Likewise, you want less important features to be pushed into the background, lower in the visual hierarchy. Thoughtful type design can accomplish this.

Hierarchy does not always follow feature size; it follows from your map purpose. A small feature, such as a tiny wetland, may be the most important feature on a map and larger features, such as surrounding towns, may be background location information. In this situation, the wetland warrants the most prominent label.

The characteristics of type that help you establish hierarchies of features are the following:

- point size
- weight
- scaling
- color lightness
- case

Point size is the most obvious type characteristic that can establish hierarchy. Larger type indicates a feature more important than those labeled with smaller type. Because size is so strongly associated with importance of a feature, you should not set a label to a larger point size simply to fill a large area. As an alternative, show large areal extents using character spacing instead. The three sizes of type shown in figure 3.19 reflect three population sizes.

Pittsburgh
State College
Boalsburg

Figure 3.19 *Differences in point size create a hierarchy of features.*

Using a bold weight for type is an obvious way to bring a feature category higher in the visual hierarchy, making it stand out from labels that are the same size and font, but that are not bold (*figure 3.20*).

Bellefonte
Pleasant Gap

Figure 3.20 *Differences in label weight also create a hierarchy.*

Horizontal scaling can be used to indicate hierarchy. The two labels shown below are the same point size and style (both are regular). Some software used in mapping may not offer a type scaling option, but "extended" and "condensed" or "narrow" fonts from a single font family can be used to distinguish the importance of features on a map. In figure 3.21, Philipsburg is extended and Port Matilda is condensed.

Philipsburg
Port Matilda

Figure 3.21 *Extended and condensed fonts can be used to indicate hierarchy.*

In figure 3.22, Centre county is larger and bold because it is a larger feature, but it is lighter, which pushes it back in the visual hierarchy. The names of smaller townships are more important for this map's purpose and remain forward in the visual hierarchy. In this case, the county category is established as a larger feature, but its visual importance is reduced with lightness.

Figure 3.22 *Lightness indicates hierarchy by pushing labels to the background.*

Altering the lightness of type can be an effective way to push a label into the background, lower in the map's visual hierarchy. To maintain the integrity of the characters when printed, light labels are often set as bold type. This is an example of a compromise that uses a type characteristic in a different way than a simple hierarchy would suggest (bolder but not more important).

Case suggests hierarchy: uppercase letters indicate a larger or more important feature than one labeled by lowercase letters. In figure 3.23, the school name is uppercase, and a building within the campus, a smaller feature that is lower in the visual hierarchy, is labeled in lowercase letters.

PENN STATE
Walker Building

Figure 3.23 *Uppercase labels are visually more important than lowercase labels.*

Size, weight, scaling, lightness, and case can be used to communicate differences in the importance of features. Importance is determined by your map purpose, not just feature size.

Ambiguity and contradiction in classification with type

Case, leading, and character spacing can be used to establish category, hierarchy, or both on a map. Uppercase lettering is often used for area feature labels regardless of feature importance because it can be spread across an area more elegantly than lowercase lettering. Thus, a group of uppercase area labels may be shown at smaller size to contradict the implied increase in importance that their case suggests.

Increasing character spacing and increasing leading pulls a label apart and diminishes its prominence in the visual hierarchy by making it less readable. This subtle push into the background happens even though these characteristics make the label larger overall.

Character spacing is also used to imply the extent of a feature: a spaced label helps show that a particular feature in a group is larger than others. Spacing also depends on the shape of the area and on the length of the feature name. Spacing has a weak effect on hierarchy because of these competing reasons for spacing lines and characters.

P E N N S Y L V A N I A
CENTRE COUNTY

Figure 3.24 *Character spacing tends to push a label back in the hierarchy because it reduces its readability.*

The following examples demonstrate how multiple type characteristics of three labels (two townships in Centre county) can be combined in ways that challenge the simple relationships described previously.

Size and lightness are used in figure 3.25 to opposite effect. Centre is simultaneously pushed both up and down in the hierarchy.

PATTON
FERGUSON
CENTRE

Figure 3.25 *Centre is larger in size but lighter than the two township labels.*

In figure 3.26, size is used with character spacing to reduce the large county label's position in the hierarchy.

PATTON
FERGUSON
C E N T R E

Figure 3.26 *In this example, the Centre label has increased character spacing, making it more difficult to read and thus offsetting its larger point size.*

Figure 3.27 uses lightness and size in another way: the background is a medium gray and labels are white and black, contrasting approximately equally with the background. This places them approximately equal in the visual hierarchy. Lightness shifts to a categorical rather than hierarchical distinction. Size has a strong influence on the hierarchy in this example, potentially pulling Centre up to greater importance than the previous example.

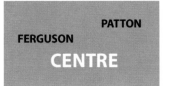

Figure 3.27 *An alternative use of lightness for labels that differ in size. All three labels are bold.*

Figure 3.28 uses three type characteristics: size, lightness, and weight. Size and weight pull Centre up and lightness pushes it back down in the hierarchy.

PATTON

FERGUSON

CENTRE

Figure 3.28 *Labels differ in size, weight, and lightness.*

Figure 3.29 uses four type characteristics: size, lightness, bold weight, and character spacing. Size and weight pull Centre up and lightness and spacing push back for fairly balanced visual importance.

PATTON

FERGUSON

C E N T R E

Figure 3.29 *Labels differ in size, weight, lightness, and character spacing.*

Size, case, weight, and lightness can be combined with character spacing and leading to separate features into groups and subtly imply differences in importance.

Label placement

Sometimes you do not need refined map labeling, particularly in dynamic displays that have multiple query options. But when sloppy labeling impairs the effectiveness of a map, how do you improve the map and make the labels easier to associate with their features? Understanding the cartographic conventions for placing the labels of point, line, and area features will help you make wise choices when arranging all the text on your map. Labeling is a time-intensive aspect of cartography, so knowledge of labeling conventions will allow you to work more effectively to produce a quality map.

Point label placement

Text that labels a point location is best positioned next to that point, horizontally aligned, and with default character spacing. Cartographers sometimes choose to curve or rotate point labels to run parallel to curved lines of latitude across the map or to arc away from coastal locations into the adjacent water body. The more care you take in curving and rotating type, the longer it takes to make a map. Use horizontal alignment for point labels unless you are committed to a craftsman approach and have a production budget that can withstand the extra time that will take.

For horizontal labels, there is a series of preferred positions relative to the point location. There is some variability in these sorts of recommendations, but the position to the right and shifted up from the point location is always rated the best. That first position puts the point closest to the base of the beginning of the label *(figure 3.30)*. Notice that label positions to the left and right of the point (1 to 4) are each shifted up or down. In figure 3.31, the "good" examples show labels in two of these positions.

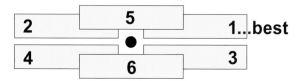

Figure 3.30 *Order of preferred positions for point location labels.*

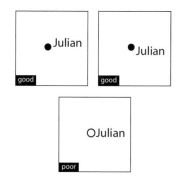

Figure 3.31 *It is better to shift labels up or down from the point rather than aligning them with the point.*

The "poor" example demonstrates the disadvantage of aligning a horizontal label directly to the right (or left) of a point location. The point symbol becomes incorporated into the shape of the word. The effect is exaggerated by using an open symbol that could be mistaken for a letter. Unless a word is completely unfamiliar, we use its shape to recognize it and read it. An aligned symbol becomes included in the shape of the word and interferes with rapid recognition of it.

Another way to ensure that the shape of a point label is easily recognized is to rely on lowercase letters. They have more distinctive features than capitals do, helping us recognize words by shape, rather than reading them letter by letter. Figure 3.32 emphasizes this by showing just the top half of the same word. With only half the information available, you can probably read the lowercase label but perhaps not the uppercase version.

Public PUBLIC

Figure 3.32 *The top half of the word "Public." Lowercase letters have more variation in shape and are therefore easier to read.*

Using a lot of uppercase lettering on a map makes that map difficult to read. Lowercase labels are not as long as corresponding uppercase labels, so they fit more easily into crowded places as well.

The recommended positions for point label placement assume that there is no other information around the point. Since that is rarely the case, label placement becomes a process of making trade-offs between guidelines and the realities imposed by the geographic distribution of features. In the example below, the two labels fit in the preferred position, to the right and up, relative to the points they label. But this places Corning equally close to both symbols. A reader can figure out which point Corning labels by a process of elimination, but you do not want to make your reader work that hard. A general rule is to position the text closer to the point it labels than to any other point with which it could be mistakenly associated.

Figure 3.33 *Unambiguous association is more important than preferred positioning for point labels.*

A set of labels dynamically positioned by ArcMap is shown in figure 3.34. Notice that they generally follow the hierarchy of preferred positions, though some are ambiguously located near more than one point. Amman is near four point symbols, Damascus is squeezed between two, and Yerevan is nearest a point it does not label.

Maintaining a consistent distance between labels and their associated points throughout the map keeps the map neat and easy to read. Figures 3.36 and 3.37 show varied gaps with the same general positions as shown in figure 3.35. This inconsistency makes for sloppy design.

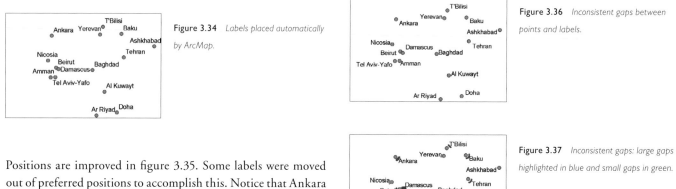

Figure 3.36 *Inconsistent gaps between points and labels.*

Figure 3.34 *Labels placed automatically by ArcMap.*

Positions are improved in figure 3.35. Some labels were moved out of preferred positions to accomplish this. Notice that Ankara was adjusted to allow a better position for Yerevan. Amman, Tel Aviv-Yafo, Beirut, and Damascus were all moved to clarify ambiguities.

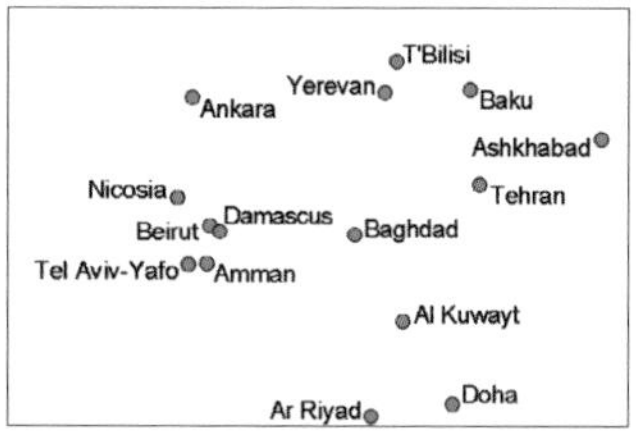

Figure 3.37 *Inconsistent gaps: large gaps highlighted in blue and small gaps in green.*

Figure 3.35 *Label positions improved with custom edits.*

Choose a consistent size for the gap based on the size of symbols, the size of the label type, and the density of labels on the map. Make this decision while viewing the map at output size, not while it is enlarged on screen.

The arrangement of other map features may also cause you to reject preferred positions for labels. The two "poor" examples in figure 3.38 show the label in the preferred position, but a curved road interferes with the label. Breaking the road for the label removes key information about the shape of the road and its relationship to the town. A better solution is to move the label to a more open position that is still near the feature.

Ideally, a point label is positioned on the same side of a linear feature as its point. In the following example, the two labels are in preferred positions, but on the opposite side of the river from the points they label. The second example shows Helvetica moved down (less preferred) and Palatino moved right. A line needs to be broken to accommodate Palatino. Choosing positions is a process of balancing conflicting guidelines.

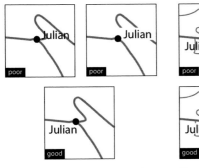

Figure 3.38 *Reposition labels to reduce interference with nearby features.*

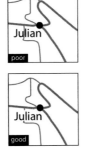

Figure 3.39 *Breaking lines for labels makes a map more readable.*

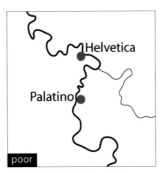

Figure 3.40 *Poorly positioned labels on the opposite side of a line feature from the point locations.*

Figure 3.41 *Labels are better positioned on the same side of the linear feature as the point location. Palatino is placed so that it breaks the most vertical part of the line feature, minimizing the amount of data masked.*

Figure 3.39 shows the most open position for the label with added roads. The small road that interferes with the label is masked to ensure that type is readable. A small portion of that road is retained above the word so that its relationship to the main road can still be seen. A general rule for label placement is to break lines for labels. Where possible, position a label such that it breaks a vertically trending line to minimize the loss of information.

When labeling point features near water bodies, position labels for coastal and shoreline features wholly in the water. Do not allow labels to span both water and land. In figure 3.42, Encinitas and Solana Beach seem to be well positioned to the right of the points they label. But these coastal places are better labeled in the water to emphasize their coastal location. Rancho Santa Fe should not run from water to land. These problems are corrected in figure 3.43.

These examples demonstrate the most important guidelines for positioning point labels:

- Position labels next to (right or left) and shifted up from the point feature.

- Maintain a consistent distance between labels and point symbols throughout the map.

- Break lines for labels, but minimize the frequency of breaks.

- Position labels on the same side of a linear feature as the point feature.

- Position labels for coastal features in water.

- Contain labels entirely on land or entirely in the water.

The task of placing point labels is often a compromise between conflicting guidelines. Placing labels so that they are unambiguous and easy to read should guide your decisions.

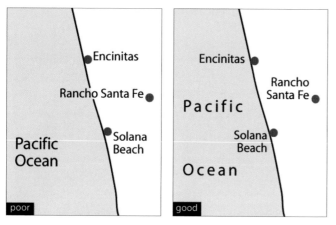

Figure 3.42 *Place labels for coastal features in the water and do not allow labels to span land-water boundaries.*

Figure 3.43 *Improved labeling of coastal and inland places.*

Line label placement

Text that labels a line feature should follow along that line and be separated from it by a small gap. The text should not have noticeable character spacing, though a moderate increase in word spacing can be useful. If a linear feature is long, repeat the label rather than adding character spacing; do not try to stretch a label along the length of the line. Figure 3.44 demonstrates many errors in label placement.

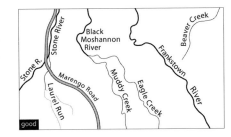

Figure 3.45 *Improved placement of linear feature labels.*

Figure 3.44 *Poorly labeled linear features.*

Several placement problems make reading this map difficult:

- □ Stone River ambiguously labels both the river and road lines.
- □ Laurel Run and Black Moshannon tip upside-down.
- □ Black Moshannon is smaller than other labels of the same importance.
- □ Marengo Road does not follow the line feature.
- □ Muddy Creek would fit better if positioned above the line.
- □ Eagle Creek is too far above the line.
- □ Frankstown River follows curves in the line too closely.
- □ Beaver Creek is touching the line.

Improvements have been made to the map:

- □ Stone River is labeled by repeating the text for the two segments of the river.
- □ Laurel Run is flipped to read right-side up. The tributary is broken for this label.
- □ Marengo Road is positioned to follow the road.
- □ Black Moshannon River is set in the same point size as other features in the same category. It is not positioned to follow the river because the name is too long to fit to the short segment of river visible on the map, though the modified left-alignment helps associate the label with the line. This is a difficult trade-off with no good solution.
- □ Muddy Creek is moved to the top side of the creek where it is a better fit to the shape of the line.
- □ Eagle Creek and Beaver Creek are positioned with gaps between the label and river that are consistent with spacing used over the whole map.
- □ Frankstown River is moved to a straighter segment of the line and fit to a smoother curve.

ArcMap and graphics software can place text along user-specified curves. With only two points, you can create a simple *c*-shaped curve or a more complex *s*-shaped curve. Rarely will you ever need three or more points to define the line on which your label will sit. The curve should be a smooth, simplified approximation of the line feature, not an exact duplicate of it. It should not cause markedly different orientations for adjacent characters. Multiple points will make a line that is cumbersome to edit and inappropriately complex. Eight-point and two-point curves are illustrated in figures 3.46, 3.47, and 3.48.

You can edit the shape of a curve by repositioning the points themselves (green squares, in the example) and by moving the handles (purple squares) attached to each point. Rotating the handles around the point, and changing their distance from it, will produce a variety of smooth curves. Experimenting by moving handles in all directions from their points is the best way to get used to altering the shape of the curve (*figure 3.49*).

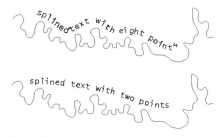

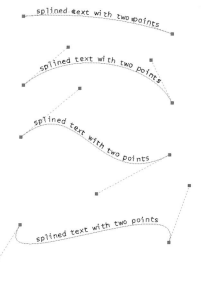

Figure 3.49 *Simple curves created by changing only the positions of two handles (purple squares).*

Figure 3.46 *Complex and simple curves used to label a line.*

Figure 3.47 *Example of overly complex splined text with eight points*

Figure 3.48 *Example of suitable splined text with two points.*

Positioning a label above a linear feature is preferable to positioning it below. Text fits best above a line because there are fewer descenders than ascenders on most labels. Also, a proper name with an initial capital letter will have a gap between the line and any lowercase letters without ascenders if placed below the line. As with other guidelines, there will be situations where you won't follow this advice to avoid overlapping other map features and labels. Regardless of the compromises you must make, avoid placing a label such that any portion of it reads upside down. Be especially aware of this when positioning vertically trending labels. Because of the shapes created by different combinations of ascenders and descenders within labels, the actual space between labels and features will vary across the map. Your challenge is to create spacing between them that looks consistent overall.

River names on the following map were placed dynamically by ArcMap. Some are well positioned; others have awkward gaps or overlaps, or they are ambiguously near more than one feature *(figure 3.50)*. Refining the software's label placement settings would resolve some of these issues.

Labels were repositioned manually in the next map *(figure 3.51)*. Labels were moved to smoother bends on the rivers, and the curves of the labels themselves were simplified. The curve of Huang He is selected to show an example of a smooth curve fit to a detailed line. Notice that Salween and Irrawaddy are oriented in opposite directions than they were previously, so neither reads upside down.

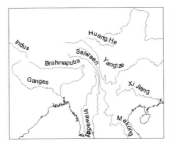

Figure 3.50 *Line labels placed automatically by ArcMap.* Source: Digital Chart of the World, ESRI.

Figure 3.51 *Manually repositioned line labels with improved positions.* Source: Digital Chart of the World, ESRI.

Positioning line feature labels according to these guidelines creates maps that are easier to read:

- Position labels to follow along portions of features.
- Use default character spacing—do not spread characters along the line.
- Repeat labels for long line features.
- Put labels at the straightest and most horizontal portion of the feature.
- Break lines that interfere with labels, but minimize the frequency of breaks by choosing label positions carefully.
- Position labels above lines when practical.
- Do not allow any part of a label to tilt upside down.
- Use simple curves for labels to fit loosely to features with complex curves.
- Maintain a small, consistent gap between labels and lines throughout the map.

Area label placement

Area feature labels indicate the extent of the feature by the way they are positioned. Area labels differ from point and line labels in that they often include character spacing. You can stretch labels using character spacing where needed, but do not express extent by making the point size of the characters larger, which instead indicates the relative importance of the feature.

Since most area features are not simple rectangles, the extent of these features can also be described using curved labels. If the shapes of these areas lend themselves to a series of simple and complementary curves, curved labels can be a pleasing addition to the design. But do not get carried away looping labels in so many directions across the map that it becomes indecipherable.

Figures 3.52 and 3.53 show an area label that uses character spacing to spread across the feature. The first label is curved to suggest both the horizontal and vertical extent of the area. The second label is horizontal and spaced to suggest extent at the widest part of the area. This label is not as fancy, but it is a reasonable (and less time-consuming) solution.

Figures 3.54 and 3.55 show the same area labeled less suitably. The first label is centered in the area and its length only partly suggests the extent of the region. The slight angle is not sufficient enough to look purposeful, so instead it looks sloppy. The second label is centered in the region without enough spread to suggest extent.

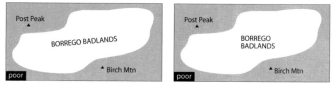

Figure 3.54 *The slight angle on the label looks like a mistake rather than an effort to label extent.*

Figure 3.55 *A centered label misses the goal of indicating the extent of the area.*

Character spacing is better suited to uppercase rather than lowercase lettering. Thus, area labels are the most likely text elements on a map to be uppercase, with other labels taking advantage of the better legibility of lowercase lettering.

Figure 3.52 *The curved and character-spaced label identifies the extent of the area.*

Figure 3.53 *Horizontal character-spaced text also makes a suitable area label.*

Figure 3.56 was dynamically labeled using ArcMap. Labels are centered and are all the same size. Figure 3.57 uses character spacing to spread labels across larger areas. The character spacing was edited manually after converting the dynamic labels to annotation. Area labels were not curved in this example.

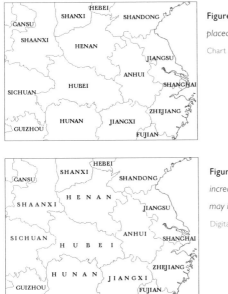

Figure 3.56 *Automatically placed area labels.* Source: Digital Chart of the World, ESRI.

Figure 3.57 *Repositioning and increasing the character spacing may improve area labels.* Source: Digital Chart of the World, ESRI.

In this map, less than a third of the labels have enough room to be character spaced. Expressing areal extent while retaining consistent label styles is a trade-off, and preference for one or the other is a subjective decision.

Labels spread over areas with character spacing should have a margin at the beginning and end of the word approximately equal to the character spacing. Do not spread words so much that the first letter crowds right against the boundary. Often some labels, like Hebei, Fujian, and Shanghai in this example, are longer than the area they label at the selected type size. Careful positioning prevents the labels from blocking characteristic parts of the area outlines. As always, break lines for labels in these cases.

In the previous example, labels are shifted up and down relative to adjacent labels. You want to avoid having a series of area labels that align horizontally, suggesting a "sentence" of labels, as seen with the poorly placed labels in figure 3.58.

Figure 3.58 *Horizontal alignments yield poor area labeling.* Source: Digital Chart of the World, ESRI.

Be aware that the letter *I* may look like a line segment if it is poorly positioned relative to area outlines. Serifs can help somewhat, but the best remedy is to shift the label so letters straddle lines and other features (large labels may also straddle smaller labels). This is particularly a problem with character-spaced area labels because letters lose their relationship to each other somewhat. One could easily read Hube instead of Hubei in figure 3.59.

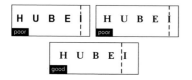

Figure 3.59 *Choose fonts and label positions to avoid ambiguous conflicts between labels and features. In this case, the letter I becomes lost in a boundary line.*

Figure 3.60 *States are labeled along the state boundaries instead of across state areas on the map. This alternative is particularly useful when only portions of areas are shown. This map shows congressional districts in a region surrounding the new Flight 93 National Memorial. Source: McKendry, Brewer, and Gardner. 2005. A Socioeconomic Atlas for Flight 93 National Memorial and its Region. National Park Service.*

It is particularly difficult to label large areas that cover much of a map, especially when only a portion of that area is shown. The letters of a conventionally character-spaced label are difficult to relate to the few visible area boundaries. One approach is to label along the boundary lines themselves, following the placement guidelines for linear features *(figure 3.60)*.

These guidelines help you create clearer maps when positioning area feature labels:

- Suggest the extent of the area by the position of the label.
- Suggest extent using character spacing and line spacing (leading).
- Suggest extent with simply curved labels.
- Use uppercase letters when spacing characters.
- Do not adjust text point size to fit labels into small areas or to fill large ones.
- Stagger horizontal alignments.
- Adjust label position so gaps fit across features.
- Ensure that individual characters are not mistaken for symbols (especially a sans serif *I* and *o*).

Dense label placement and trade-offs between rules

Leader lines can assist in labeling very crowded map areas. They improve clarity by allowing map labels to be placed further from their features while retaining an association with them. Use as few leader lines as possible; consider them a last resort in label placement. Do not put a leader on every label, as shown in the whimsical example below.

Figure 3.62   *Over-reliance on breaking lines for labels obscures important map features.*

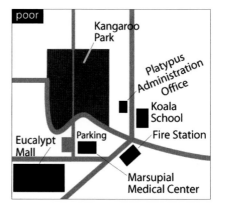

Figure 3.61   *Over-reliance on leader lines results in an overly complicated map.*

Figure 3.63   *Redesign of some labels and use of one leader line clarifies the map.*

The same map is labeled again with no leader lines, but multiple breaks in roads are made, including a major intersection to accommodate Marsupial Medical Center. Important relationships among features have been lost (*figure 3.62*).

A third solution uses just a single leader line and breaks only straight portions of roads. It moves the park and mall labels inside their features, appropriately treating them as areas. This option offers the clearest associations between labels and features without compromising the relationships between features (*figure 3.63*).

The following examples demonstrate how editing dynamically placed labels can improve a map. Parameters were set to position point labels above and to the right of point features, and to put line labels above the line features. Overlaps were allowed to ensure that the maximum number of labels were placed. Type characteristics were also set as parameters prior to dynamic labeling to further distinguish between roads and landmark buildings. A type size large enough to accommodate map readers with poor vision was used *(figure 3.64)*.

The dynamic labels were converted to annotation and repositioned to remove overlaps and clarify ambiguities. Adjustments were made, label by label, to improve associations between labels and features. The labels of three of the major streets were changed to bold type to improve feature classification *(figure 3.65)*.

The large type and small size of the map made fitting a complete set of labels onto this map a challenge. After experimenting with various positions, just a single leader line to the library was necessary to label all roads and landmarks, and to resolve overlaps. Notice that few of the point labels remained in the preferred position—up and to the right—after all the constraints on label positioning were considered.

Refining the positions of labels requires experimentation. It depends on type sizes and styles, so set characteristics carefully before you begin the process of dynamic labeling. Custom editing is an important, yet time-consuming, part of the design process. Use leader lines sparingly to make key links only in the densest areas of a map.

Figure 3.64 *Automatically placed and symbolized labels in ArcMap.*

Source: ESRI ArcGIS Sample Maps data.

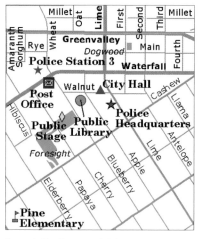

Figure 3.65 *Annotation labels were repositioned to clarify links to features.*

A single leader line was used for the tightest space, associating the public library label with its symbol. Source: ESRI ArcGIS Sample Maps data.

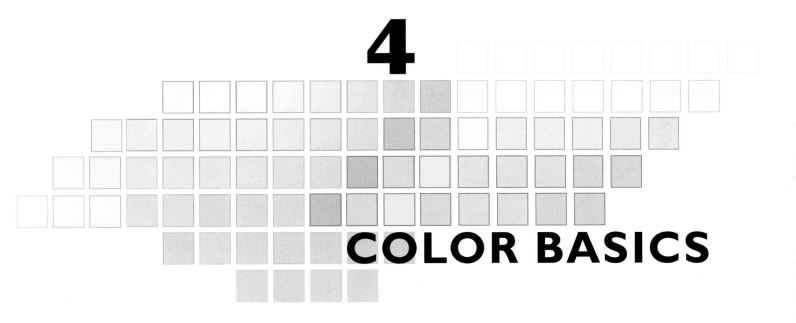

4

COLOR BASICS

CHAPTER 4

COLOR BASICS

Most map designers who work with digital data create maps in color on their computer screens. But choosing from among the millions of colors that can be specified for the screen is a daunting task. If all you had to do was chose pretty colors, the task would be fairly straightforward; but I fear too many maps would end up blue, the most common favorite color.

Choosing map colors goes beyond considering what colors might be related to the mapped topic, such as blues for bodies of water and browns for pollution data. Since people's opinions about which colors they like and which colors best represent a topic often conflict, this chapter takes an analytical approach to choosing map colors.

GIS and graphics software typically offer three color systems for specifying color on maps: HSV, CMYK, and RGB. Why learn more than one system? HSV makes partial use of perceptual dimensions, so it can be used to adjust the appearance of map colors. CMYK is the language of graphic arts, so if you are printing your maps and adjusting colors for print, you need to be conversant in this system. RGB is the main color system used for computer graphics, so designing for screen display may require you to switch to this method of color specification.

You will probably choose only one of these three systems for most of your work, since the software can make conversions between them for you. Expert color communication, however, will require you to understand color specification in all of the systems to produce the color characteristics suited to each mapping challenge. To do this, you will need to understand the utility of three-dimensional perceptual color space and specify colors with specific hue, lightness, and saturation characteristics in HSV, CMYK, and RGB systems.

Map designers who produce quality maps know how to work with color on their computers, considering the following:

▫ perceptual dimensions of hue, lightness, and saturation

▫ perceptual color systems and their relationship to HSV and color mixture cubes

▫ how to mix color to create map symbols using CMYK and RGB

Perceptual dimensions

Maps mostly fall into two categories: general reference maps and thematic maps. On general reference maps, color hue is used to symbolize different kinds of features. A simple example is using blue and green to differentiate water from land areas, but often maps show features that people do not see on the ground and that have no direct association with particular hues. When mapping this sort of thematic data, hue has an analytical purpose. You will rarely choose colors that are associated with land surface characteristics when designing a thematic map. For example, public land may be mapped in yellow and private land in brown. Areas of a state that lost population could be shown in orange, and areas that gained population could be shown in purple.

The blue/green, yellow/brown, and orange/purple examples seem to simply be pairs of "colors," so why use this extra word, "hue?" When using color as a symbol, each color is a combination of three perceptual dimensions: hue, lightness, and saturation. Hue is the most familiar and, therefore, most easily understood dimension. Lightness is the most important dimension for representing quantitative data. Saturation has the most subtle use in map symbology, but uncontrolled saturation differences can wreak havoc with a map's effectiveness.

Color is three dimensional. You can split colors into hue, lightness, and saturation dimensions. By mixing red-green-blue or cyan-magenta-yellow primaries, you can produce and adjust colors with the perceptual attributes you want to see on your maps. The overall goal is to envision the map colors you want and then be able to create them with any of the color mixing tools offered by GIS and graphics software, such as RGB, CMYK, and HSV.

Hue

Hue is the perceptual dimension of color that we associate with color names, such as red and yellow. (Dominant wavelength is a similar measure used in physics.) The rainbow, or spectrum, places saturated hues in wavelength order from long-wavelength reds to short-wavelength blues: red, orange, yellow, green, blue. Purples and magentas are not colors of the rainbow. They result from mixing red and blue light from opposite ends of the spectrum *(figure 4.1, opposite)*.

White sunlight contains a full range of visible wavelengths, but our television and computer screens get by with a reduced set of primaries—red, green, and blue (RGB) light—which are mixed to form all the other hues. Cyan, magenta, and yellow (CMY) pigments are used to mix colors for printing and painting. Mixing lights is called additive mixture (RGB are the additive primaries), and mixing pigments is called subtractive mixture (CMY are the subtractive primaries) *(figures 4.2 and 4.3)*.

Figure 4.2 *Red, green, and blue are the primary hues used to mix colors with light on computer screens.*

Figure 4.3 *Cyan, magenta, and yellow are the primary hues used to mix colors with ink.*

The hue circle in figure 4.4 is constructed using these two sets of primaries: RGB and CMY. Notice that magenta lies between red and blue, the ends of the spectrum.

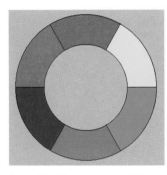

Figure 4.1 *Set of ten hues shown in spectral order.*

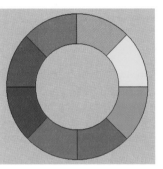

Figure 4.4 *A hue circle showing RGB and CMY primaries in spectral order.*

Figure 4.5 *This hue circle shows unique hues in opponent positions across the color circle from each other—red across from green and yellow across from blue. Notice that this hue circle differs from the previous example built from primaries. There is no single correct hue circle, but each shows hues in spectral order.*

You may be thinking, "But I learned that red, yellow, and blue were the primary colors when I mixed paints as a kid." Please set aside your school days red-yellow-blue approach to color. It may confuse you in this modern era of computer screens and color laser printing.

The red-yellow-blue (RYB) color set, however, is akin to the perceptual concept of unique hues. The unique hues that do not look like mixtures of other hues are red, green, yellow, and blue. You may find yourself gravitating toward these unique hues (RGYB) as you develop color codings because they are clearly different from each other. Red-green and yellow-blue are also the opponent hues on which our eye-brain color vision system is based. In addition to red, green, yellow, and blue, the other basic colors named in all fully developed languages are pink, purple, orange, brown, gray, white, and black.

You may notice that the more unfamiliar colors mentioned, magenta and cyan, are not in the list of common color names. You have probably encountered both of these hues when working with color desktop printers and color photocopiers or if you have communicated with publishers. Magenta is a pinkish red hue (a red with no yellow) and cyan is a greenish blue (a blue with no red). The purer these primaries, the more varied the hues that can be mixed with CMY.

The hues shown in the hue circles on this page vary in lightness. Figure 4.6 shows the effect of holding lightness constant and changing only hue.

Figure 4.6 *Five hues shown at the same lightness.*

Four light hues are shown on the map *(figure 4.7)* to differentiate telephone exchange areas. Hues are used on this map in the simplest way to distinguish one area from another. The hues are not used as symbols. That is, they do not indicate characteristics of the exchanges, such as number of calls or differences in companies providing local exchange telephone service.

For figure 4.7, any area can be any of the colors used as long as two areas with the same color do not share a boundary. Only four hues are needed to solve this type of map coloring problem (called the four color theorem in mathematics), but this strategy has limited application in cartography because color is not being used to represent attributes of the areas.

A second map of telephone exchanges *(figure 4.8)* uses hue as a symbol to indicate each of the companies providing telephone service. There are eight companies serving the mapped area, so eight hues are needed to represent this dataset. The three areas represented in pink indicate three areas served by the same local exchange carrier (Granite State Tel. Co.). Pink cannot be used elsewhere on the map for exchanges not served by that carrier.

Differences in hue indicate different kinds of features. For example, hue is used differently on the two maps in figures 4.7 and 4.8. It differentiates areas in the first example and indicates area attributes in the second.

Figure 4.7 *Telephone exchange areas around Concord, New Hampshire. Four hues are used to differentiate exchange areas.* Source: TeleAtlas.

Figure 4.8 *Local exchange carriers are mapped for each exchange area. Eight hues are used to represent the eight companies that provide telephone service to Concord, New Hampshire, and its surrounding region.* Source: TeleAtlas.

Lightness

Variation in lightness is frequently used to represent a ranking within mapped data. Light colors are usually associated with low data values and dark colors with high data values. For example, shallow water may be represented by light blue and deep water by dark blue. Areas with high death rates may be represented by dark red and areas with low death rates by light yellow.

The map in figure 4.9 includes a hue change (yellow-orange-red) as well as a lightness change (light to dark), but it is the lightness change that most clearly communicates the low-to-high sequence in the data.

Changing lightness only, while maintaining a constant hue, produces the map shown in figure 4.10. Colors range from dark to light red. Compared to the previous map, this map has less contrast between adjacent colors because the scheme has not been augmented by variations in hue.

Lightness is a relative measure *(figure 4.11)*. It describes how much light appears to reflect (or emit) from an object compared to what looks white in the scene. Its relative character makes it a different measure from related terms like brightness, luminance, and intensity. Another word for lightness is "value," but that term becomes confusing in quantitative work if you are also describing data values.

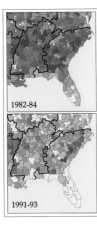

Change in Stroke Deaths

White Males Southeast U.S.

Comparison of 1982-84 and 1991-93 Rates

1982-84

Death rate per 100,000 white males by health service area

| 54.0 to 91.5 |
| 43.8 to 54.0 |
| 37.1 to 43.8 |
| 31.8 to 37.1 |
| 26.6 to 31.8 |
| 19.8 to 26.6 |
| 4.0 to 19.8 |

1991-93

Source: National Center for Health Statistics

Figure 4.9 *Stroke rates for white males over two time periods are represented with colors that range from light for low rates to dark for high rates. A hue transition, from yellow to red, is used to augment color differences, but lightness carries the main message of the map.* Source: National Center for Health Statistics.

Figure 4.11 *Four blues that vary in lightness from light to dark.*

All of the colors in the lightness sequence *(figure 4.12)* share the same orange hue. Notice that brown, at the end of the sequence, is actually just a dark and desaturated orange. This final perceptual dimension of color, saturation, is the most conceptually difficult to understand.

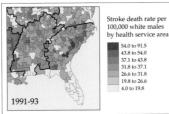

Stroke death rate per 100,000 white males by health service area

| 54.0 to 91.5 |
| 43.8 to 54.0 |
| 37.1 to 43.8 |
| 31.8 to 37.1 |
| 26.6 to 31.8 |
| 19.8 to 26.6 |
| 4.0 to 19.8 |

1991-93

Figure 4.10 *Stroke rates are shown with lightness differences only.* Source: National Center for Health Statistics.

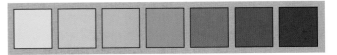

Figure 4.12 *A series of seven colors of medium saturation that change gradually in lightness.*

Saturation

Saturation is a measure of the vividness of a color, and there are a host of terms related to saturation with slightly varied definitions: chroma, colorfulness, purity, and intensity (although intensity is confusingly used to describe both lightness and saturation). Some of these terms are from perception and the scientific realm of psychophysics, and some describe the physics of light. Still others, like shades, tints, and tones, are from art. As any hue is desaturated, it becomes more gray, a neutral color with no hue.

You can use saturation changes to reinforce lightness changes to help distinguish map symbols. For example, you may choose colors that range from light-desaturated to dark-saturated. This strategy is similar to using hue changes to augment lightness differences, as seen in the yellow-orange-red map showing stroke death rates (*figure 4.9*). Saturation can also be used with variations in hue and lightness to add emphasis to categories with small polygons.

Figure 4.13 *A set of dark purples ranging from near gray to more vivid purple that demonstrate a saturation range while retaining the same hue and same lightness.*

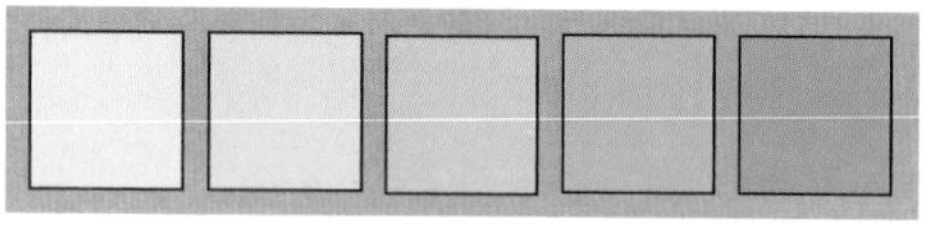

Figure 4.14 *A set of colors with decreasing lightness, ranging from desaturated gray to saturated cyan.*

Only a few noticeably different steps will be available when you vary just the saturation (and not the hue or lightness) of a symbol. Therefore, using saturation as the primary difference among symbols is difficult. Figures 4.15 and 4.16 show elevation data with saturation steps (greens) and with lightness steps (grays). Note how poorly the landforms are presented by varying saturation alone.

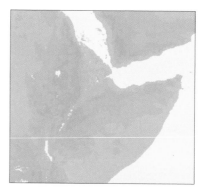

Figure 4.15 *Saturated yellows represent higher elevations, and desaturated grays represent the lowest elevations in the Horn of Africa region. These colors have constant hue and lightness. Saturation alone is not an effective symbol variable.* Source: USGS.

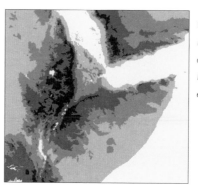

Figure 4.16 *A series of grays from light to dark represent the range of elevations for the Horn of Africa. Lightness alone represents these elevation ranges well.* Source: USGS.

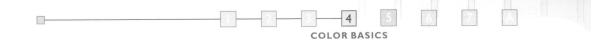
Notice the difference in the relationship between lightness and saturation in the two examples below (figures 4.17 and 4.18). Both would be useful in mapmaking because the perceptual dimensions are used systematically.

Figure 4.17 *A series of four colors with the same hue, decreasing in lightness and increasing in saturation. They range from light-desaturated red to dark-saturated red.*

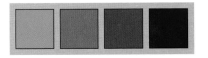

Figure 4.18 *A series of four colors with both lightness and saturation decreasing. This set ranges from light-saturated red to dark-desaturated red.*

Figure 4.19 *Two sets of five hues, all with the same lightness. The top set all have the same saturation. In the bottom set, yellow has a much higher saturation than the other hues. If these colors were used on a map, the data represented with yellow should be highlighted for a reason. It should be either an important category or a tiny area that would otherwise be lost.*

Saturation is the hardest of the three color dimensions to use in map design. Even if you do not use it explicitly, ignoring saturation can produce a confusing map. Individual colors that are accidentally more vivid than others will stand out strongly from other symbols for no apparent reason. These lapses will put inappropriate emphasis on map categories that are not more important than others. Be particularly careful about choosing colors for large map areas; large areas of high-saturation colors will dominate the look of the map.

The map legend "Major Habitat Types," might be seen as an unsuitable mix of saturation levels. The yellow, red, and purple are more saturated than the other colors. They are also either much lighter or darker than other colors. But consider the map—the highly saturated colors are used for very small areas. The map works because saturation makes these small features visible.

Figure 4.20 *A map of major habitat types in central Mexico and its legend purposely use both saturated and desaturated colors. Grasslands and mangrove habitats, which cover very small areas, are represented with high-saturation colors that are also lighter or darker than the majority of the map. Saturation is used appropriately to emphasize these very small areas on the map.* Source: ESRI ArcGIS Sample Maps data.

Perceptual color systems

The perceptual dimensions of color can be used to construct three-dimensional color spaces. Selecting colors from a true perceptual space would be the ideal tool for designing color schemes for maps, but unfortunately, the ability to match your own mental image of a complete three-dimensional perceptual color space is not yet available in GIS or graphics software. The HSV color system that is offered in some GIS and graphics software is a poor approximation.

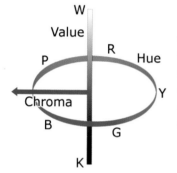

Figure 4.21 *The Munsell color space is a color order system that is perceptually scaled. Its perceptual dimensions are hue (arranged radially around the value axis), value (a vertical lightness dimension), and chroma (increasing outward from the value axis).* Source: Munsell color system.

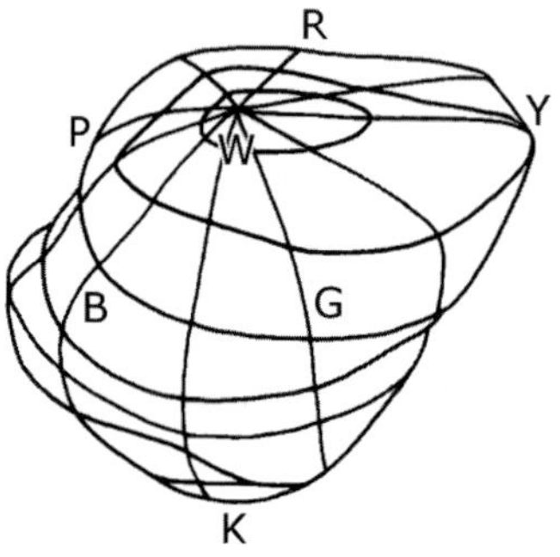

Figure 4.22 *Filling the volume of Munsell color space produces an oddly shaped, asymmetrical solid. This lumpy shape happens because high-saturation (high-chroma) yellows are light (the upper part of the color solid bulges toward yellow) and high-chroma purple blues are dark (the lower part of the solid bulges toward purple and blue).* Source: Munsell color system.

Three-dimensional color spaces

Different systems for specifying and representing color use different dimensions. RGB and CMY both have three dimensions, but they mix color in different media (light versus inks). Along with these two systems, ArcMap also lets you mix color using hue-saturation-value (HSV). Similarly, hue-saturation-brightness (HSB) color mixing is offered by Windows operating systems, and other perceptual color systems include HLS, HVC, Ljg, Luv, and IHS. All have three dimensions that can be used to conceptualize three-dimensional spaces, which are useful for structuring the way we think about and use color.

Some of the most rigorously designed color systems from color science are CIELAB (a system with L,a,b dimensions defined by the Commission Internationale de l'Eclairage), Munsell® (HVC—hue, value, chroma—*see figures 4.21 and 4.22*), and the Optical Society of America Uniform Color Scales (OSA–UCS Ljg). All of these systems arrange hues in spectral order around a central vertical axis of lightness. Saturation or a similar measure increases outward from the lightness axis.

Each of these system claims to be at least partly perceptually scaled, such that equal distances in color space produce equal color difference perceptions. For example, if a red and a purple red are ten units apart, two blues that are also ten units apart should appear equally different.

A vertical slice through the yellow and blue hues of the Munsell color space solid is shown in figure 4.23. The colors increase in saturation (chroma) away from the central axis of neutral grays, which is shown between the two vertical black lines. Inherent differences in lightness between the two hues cause the shape of the purple and blue "hue leaf" on the left to be quite different from that of yellow on the right. A useful set of map colors *(figure 4.24)* can be selected using a systematic path through color space *(figure 4.25)*.

Computer science offers some poorer cousins to these perceptual spaces that are simpler to include in software interfaces. Two common ones, hue-saturation-value (HSV) and hue-saturation-brightness (HSB), seem to be perceptual systems because they make use of the hue, brightness/value, and saturation terminology but are actually straightforward mathematical transformations of RGB. They can be useful for color selection, but you should be critical of them as you use them.

Figure 4.24 *A color scheme selected from the arc across the pair of hue leaves shown below in figure 4.25.*

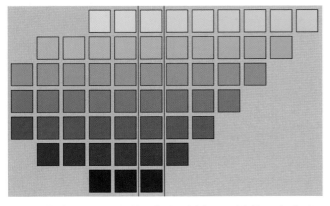

Figure 4.23 *A slice through the Munsell color solid along purplish blue and yellow hues. Notice the different shapes of each hue leaf when colors are ordered perceptually.* Source: Munsell color system.

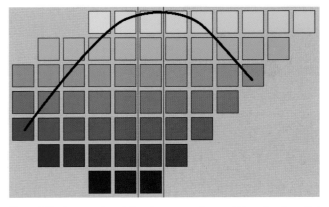

Figure 4.25 *A systematic path through a perceptual space yields a color scheme that would be useful for mapping thematic data. An arc is sketched through blues that get lighter and more desaturated and then shifts to yellows that increase in saturation and get darker.* Source: Munsell color system.

HSV: Hue-saturation-value

The HSV color system is a symmetrical cone-shaped color space. It is handy to have access to approximate perceptual specifications for creating map colors, but true color perception does not produce a symmetrical color space. Therefore, the symmetry of HSV produces flaws in its perceptual characteristics that you need to work around *(figures 4.26 and 4.27)*.

All saturated hues are on the upper and outer edge of the HSV cone, regardless of their intrinsic differences in lightness. This symmetry means that value specifications between hues are not comparable. For example, a saturated yellow and saturated blue are designated as the same "value" but have wide differences in perceived lightness.

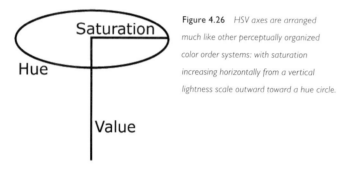

Figure 4.26 *HSV axes are arranged much like other perceptually organized color order systems: with saturation increasing horizontally from a vertical lightness scale outward toward a hue circle.*

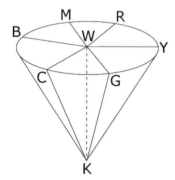

Figure 4.27 *The HSV color solid is a regular cone. Contrast this symmetrical shape with the irregular form of the perceptually ordered Munsell solid (figure 4.22).*

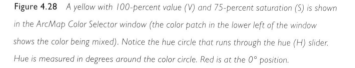

Figure 4.28 *A yellow with 100-percent value (V) and 75-percent saturation (S) is shown in the ArcMap Color Selector window (the color patch in the lower left of the window shows the color being mixed). Notice the hue circle that runs through the hue (H) slider. Hue is measured in degrees around the color circle. Red is at the 0° position.*

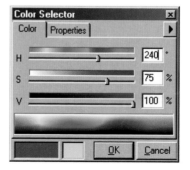

Figure 4.29 *This blue has the same value and saturation settings as the yellow above, but it is much darker.*

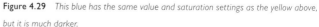

If you want to use the HSV system to select a set of different hues that have similar values, you need to do it by eye rather than using the V units. This is because the top of the HSV cone is flattened, with white pressed down to the same lightness as the pure hues. The flat top causes all saturated hues to have the same value. For example, at high value, low to high saturation intervals progress from white to green. Green is darker than white, so lightness and saturation have been combined, making both difficult to control.

The symmetry of the HSV system makes it difficult to control the lightness of colors in a systematic manner when using multiple hues or when adjusting saturation. Develop your ability to see the three perceptual components of colors and then use this knowledge to work around the flaws of HSV. In situations where much tweaking is required to achieve the desired effect, the system offers little benefit over grappling with raw specifications in RGB or CMY.

Figure 4.30 *For this set of three colors, hue is constant, value is maximized at 100 percent, and saturations are set at 0, 25, and 50 percent, respectively. Notice that the lightness of these three colors varies, showing how lightness and saturation are confounded in HSV. Zero saturation combined with high value settings yields white.*

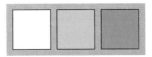

Figure 4.31 *HSV does a better job of maintaining lightness at lower value settings. The three colors here each share the same hue, have a value of 50 percent, and saturations of 0, 50, and 100 percent, respectively. At lower value settings, zero saturation yields a more accurate gray. The middle color was set in the Color Selector window shown below.*

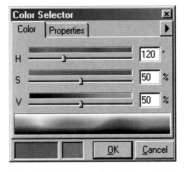

Figure 4.32 *At lower values, lightness is held more constant within the saturation dimension. Notice the improved consistency along the saturation (S) slider in the Color Selector window.*

Color cubes

Although we want to control the perceptual dimensions of color, we often must work with color mixtures of CMY and RGB because they specify the amount of pigment or light that produces map colors. The color spaces for mixing primaries are not perceptually scaled like HVC or Lab. Instead, mixtures of the three primaries of either system form a regular 3D cube. The color cubes are like three-dimensional graphs with three axes (*figure 4.33*).

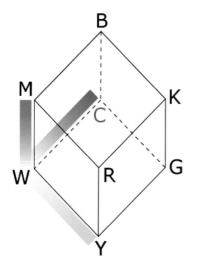

Figure 4.33 *CMY color mixtures produce a cube-shaped space. K is used as the abbreviation for blacK, since B is used for blue.*

The CMY cube has a cyan (C) axis, a magenta (M) axis, and a yellow (Y) axis. White is at the origin of these axes (zero percent of each primary). Full amounts of each primary mix to black (indicated by *K* for blacK) at the opposite corner of the cube from white. Look to each corner to see what colors (secondary hues) are made by mixing pairs of primaries. Cyan and magenta mix to blue. Magenta and yellow mix to red. Yellow and cyan mix to green.

Notice that all of the subtractive primaries are light colors, their combinations in pairs mix to darker colors, and finally all three make the darkest color of all, black (*figure 4.34*). Did you also notice that the secondary hues mixed from the subtractive primaries are actually the familiar RGB set?

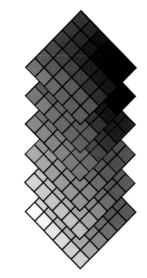

Figure 4.34 *The slices through a color cube shown here are built from mixtures of six steps in each primary. The cube is pulled open along the magenta axis to demonstrate the volume of colors it encloses.*

The RGB color cube looks exactly the same as the CMY cube, but the color mixing relationships are reversed *(figure 4.35)*. The origin of the RGB axes is black (no light). Each pair of primaries mix to the lighter CMY secondary colors, and full amounts of all three mix to white. Understanding that colors get lighter as amounts of the additive primaries are increased is a key to learning how to mix colors with RGB.

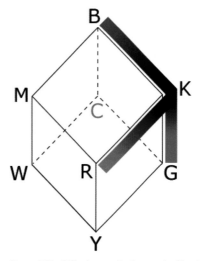

Figure 4.35 *RGB mixtures also form a cube. The similarity between the CMY and RGB cubes is emphasized here by retaining the orientation of the CMY cube.*

These color cubes that describe color mixing are more closely related to perceptual color space than expected. For example, lightness steps are arranged diagonally through the cube. You can tip the cube in figure 4.36, with the sequence from white through gray to black arranged vertically, for a rough approximation of perceptual color space *(figures 4.36 and 4.37)*.

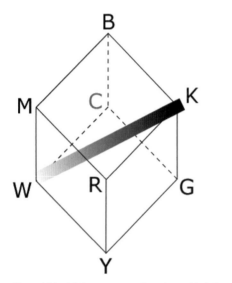

Figure 4.36 *A lightness axis runs from white to black diagonally through the color cube.*

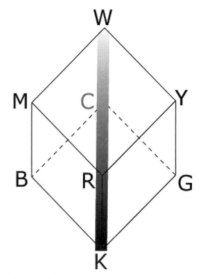

Figure 4.37 *Tip the color mixture cube on its point to see the similarity between this space and the perceptual spaces discussed earlier. The lightness axis runs vertically through the cube in this orientation.*

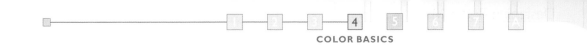
In addition, the lighter subtractive primaries are positioned above the darker, additive primaries. The hues are arranged around the cube in spectral order, like a hue circle. Each of these characteristics is consistent with the organization of perceptual color spaces *(figure 4.39),* but RGB and CMY specifications describe primary mixtures rather than perceptual characteristics. You will see the links between perceptual dimensions and color mixing in the guidelines for mixing color in the next section.

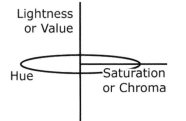

Figure 4.38 *A generic structure for perceptual dimensions of color includes a vertical axis for lightness, a horizontal axis for saturation, and a circular arrangement of hues in spectral order. This structure is shared by most color-order systems.*

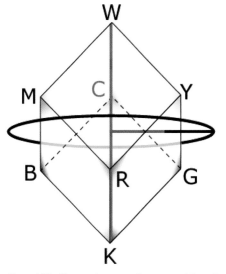

Figure 4.39 *The generic structure for perceptual dimensions fits the tipped color-mixture cube reasonably well.*

How to mix color

In computing environments, you use one of two sets of primary colors to mix other colors. On screen, you mix tiny dots of red, green, and blue light. On paper you mix tiny dots of cyan, magenta, and yellow inks. These systems work in opposite ways.

Because printing inks lack the purity of light, there are discrepancies in the range of colors that can be mixed in the two systems. You cannot mix all of the vivid colors of RGB light emitted from your computer screen using CMY printing inks on the page. Software programs can translate between CMY and RGB, though the results are not always as expected. You can visualize the difference by picturing the RGB cube as being a little bigger than the CMY cube, even though they look generally the same.

General guidelines

The relationship between the RGB and CMY color cubes can be further simplified. Looking down on the color cube from the white corner, notice that the hues at the corners of the cube are arranged in the order of the colors of the spectrum (*figure 4.40*). This arrangement is a lot like the hue circle.

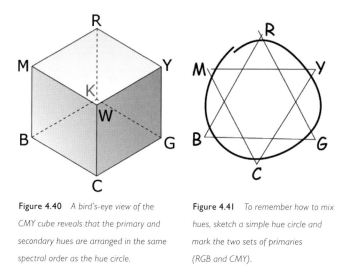

Figure 4.40 *A bird's-eye view of the CMY cube reveals that the primary and secondary hues are arranged in the same spectral order as the hue circle.*

Figure 4.41 *To remember how to mix hues, sketch a simple hue circle and mark the two sets of primaries (RGB and CMY).*

We can simplify this order of hues to a hue circle with six primary and secondary hues: red, yellow, green, cyan, blue, magenta, and back to red (*figure 4.41*). Practice sketching this simple hue circle; it will help you remember how to mix hues for both color mixture systems.

If you are mixing CMY inks, then CMY are the primary colors and RGB are the secondaries. Notice that the red secondary color falls between yellow and magenta primaries on the circle; equal amounts of Y and M mix to R. More Y than M mixes a yellow red (orange).

Green falls between Y and C primaries on the circle, so equal amounts of each mix green. More C than Y mixes a bluish green. Mix hues by choosing proportions of adjacent primaries.

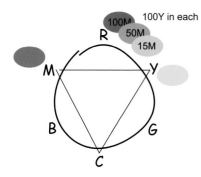

Figure 4.42 *Red is between the yellow and magenta ink primaries on the hue circle. Equal amounts of magenta and yellow mix red. Reducing the amount of magenta moves the color mixture closer to yellow through oranges.*

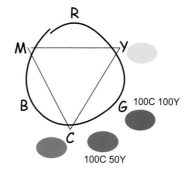

Figure 4.43 *Green is between the cyan and yellow primaries on the hue circle. Equal amounts of cyan and yellow mix green, and more cyan than yellow moves the mixture around the circle to blue green.*

When mixing hues with RGB, the circle is used in the same way, but RGB are the primaries and CMY are the secondaries. Yellow falls between red (R) and green (G), so equal amounts of R and G light mix to yellow (Y). Even though this may be an unfamiliar way to mix yellow, you can remember it if you can recall this simple hue circle. Equal amounts of blue and red mix to magenta, and more B than R mixes a more bluish purple hue.

CMY mixtures are specified with percentages of ink, which tells you how much of the page is covered by a thin film of ink broken into tiny dots. RGB mixtures are usually specified using numbers from 0 to 255 because that range balances detail and efficient computation. The emphasis in this chapter is on general amounts, more or less ink, and the perceptual results they produce, but you will see the different measurement scales reflected in examples.

The same hue circle helps you remember how to mix hues with either set of primaries, so all that remains is to adjust lightness and saturation. Lightness works in opposite ways in the two systems. Higher amounts of CMY in subtractive mixtures produce darker colors. For example, 100 percent M is darker than 20 percent M. In the example below, a light green is mixed using low percentages of Y and C. Conversely, higher amounts of RGB in additive mixture produce lighter colors. In RGB, 255 R is lighter than 50 R, and dark colors are mixed by reducing amounts of the primaries.

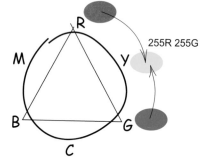

255R 255G

Figure 4.44 *The color circle works just as well for remembering how to mix hues with RGB. The most surprising mixture is yellow from red and green primaries.*

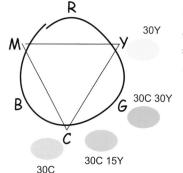

30Y

30C 30Y

30C 15Y

30C

Figure 4.45 *Combinations of lighter (low percentage) cyan and yellow subtractive primaries mix to produce light greens.*

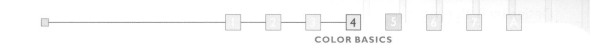
Saturation is the one color dimension that you can control with a similar strategy in both systems. Big differences in primary amounts produce more saturated colors in both systems, while nearly equal amounts of the three primaries produce less saturated colors in both systems. For example, high M and no C or Y makes a saturated magenta in CMY mixing. A desaturated magenta is produced using similar amounts of C, M, and Y, but with the M only a little higher than the other two, as shown in figure 4.46.

A saturated magenta is produced with an RGB mixture of high R and B, and no G. A big difference between one primary amount relative to the other two is the key to high saturation. A desaturated magenta is produced by mixing similar amounts of RGB, but with G (magenta's complement) a little lower than the other two *(figure 4.47)*.

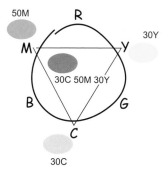

Figure 4.46 *Magenta (M) is desaturated by adding small amounts of the other two subtractive primaries, cyan (C) and yellow (Y). Remember that this view is looking down on the perceptual dimensions structure, so saturation decreases toward the center of the circle. The color is lightened by reducing percentages of all of the primaries.*

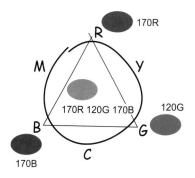

Figure 4.47 *Magenta (equal R and B) is desaturated by including a lower amount of the third primary, G. The color is darkened by reducing amounts of all of the primaries. Combining low numbers of the additive primaries, RGB, produces a darker color—the opposite result of combining low percentages of the subtractive primaries, CMY.*

CMYK mixing

Although color systems are defined by only three dimensions, subtractive color mixture often uses four inks, rather than just the three subtractive primaries, cyan, magenta, and yellow. All hues—as well as black—could be mixed using only cyan, magenta, and yellow ink. But since black is such a common color in print, printers use black ink to clearly render black text and lines and to print grays and dark hues more accurately. In offset lithographic printing—the method used to print most color publications—printing with CMYK inks is referred to as four-color process printing.

CMYK percentages range from 0 percent (no ink) to 100 percent for full coverage of ink. A specification of 20 percent, for example, indicates that 20 percent of the colored area will be covered with tiny dots of solid ink and 80 percent of the white paper will show through, producing a light color. Following are detailed guidelines for subtractive color mixture.

1. Set hue using a single ink (C, M, or Y alone) or using proportions of two of these colored inks.

Figure 4.48 *20C 40Y (left) is approximately the same hue as 50C 100Y (right). There is half as much cyan as yellow in both color patches.*

2. Set lightness using the overall magnitude of C, M, and K. Higher percentages produce darker colors. Y will remain light regardless of its percentage, so it has minimal affect on lightness.

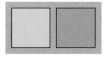

Figure 4.49 *15M 5K (left) is lighter than 30M 15K (right). Both magenta and black darken the color.*

Figure 4.50 *Both colors contain 50M. The patch on the right also has 30Y, which changes the hue but does not darken the color.*

3. Set saturation by adding K or by adjusting the primary with the smallest magnitude.

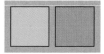

Figure 4.51 *30C 30Y (left) is more saturated than 30C 30Y 15K (right). Black desaturates the color.*

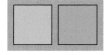

Figure 4.52 *30C 30Y is also more saturated than 30C 15M 30Y (right). The color on the right is desaturated with magenta ink. It is a warmer desaturated color than the one made with black in figure 4.51.*

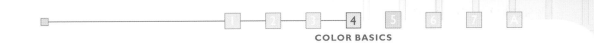
4. Create systematic perceptual changes by making systematic percentage changes.

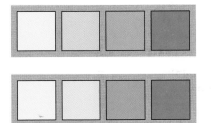

Figure 4.53 *Progressive magnitudes of 10, 30, 60, and 100 percent cyan (top series) look more evenly spaced than a series of 10, 20, 70, and 100 percent cyan (bottom series). Notice the large gap between 20 and 70 in the bottom series.*

5. Equal percentage steps do not look like equal visual steps. Use bigger steps in higher percentages.

Figure 4.54 *A magnitude step of 5 to 20 percent black (top pair) looks more different than 80 to 95 percent black (bottom pair), even though they both have a 15 percent difference.*

6. Do not use all four inks at once. Desaturate or darken with either K or the least percentage of CMY.

Figure 4.55 *Three inks (10C 40M 80Y on left) are easier to work with than all four inks (5C 40M 80Y 5K on right). Both colors look the same.*

The second way of desaturating and darkening colors by using colored inks (CMY—*figure 4.52*) is important for printing some maps because map colors are then easier to adjust and control on press. The press operator's adjustments to ink flow for high-quality black type and lines will not overly darken and desaturate colors if they are not dependent on the black component.

Darkening and desaturating with black has advantages too. The hues in subtle color designs are more stable at the press and in varied media if they don't contain all three CMY inks. For example, a beige won't shift toward blue if black rather than cyan is the third ink that desaturates orange to beige.

RGB mixing

RGB color mixtures are usually specified with numbers that range from 0 to 255. Below are detailed guidelines for additive color mixture (RGB color can only be approximated with printing inks for these example figures):

1. Set hue using one or two RGB primaries. When hues are created using two primaries, similar proportions produce similar hues.

Figure 4.56   *100R 50B (left) is approximately the same hue as 240R 120B (right). Both have half as much blue as red.*

2. Set lightness using the overall magnitude of RGB numbers. Higher RGB numbers produce lighter colors.

Figure 4.57   *200B (left) is lighter than 100B (right). The lighter color has a higher amount of the blue primary in it.*

3. Set saturation using the lowest RGB number.

Figure 4.58   *100R 200G 230B (left) is more saturated than 170R 200G 230B (right). Notice that different amounts of red, the lowest of the three primaries, produce the saturation difference.*

4. Create systematic perceptual changes by making systematic RGB changes.

Figure 4.59   *The first series of four colors (100R, 150R, 200R, 255R; top) looks more evenly spaced than the second series (100R, 130R, 225R, 255R; bottom), which includes a large gap between 130 and 225.*

5. Equal steps in RGB numbers do not look like equal visual steps. Use larger steps in the lower magnitudes to differentiate between dark colors.

Figure 4.60   *200G and 255G (left and right in top pair) look more different than 0G and 55G (bottom pair), though both pairs have differences of 55.*

Red, green, and blue mix together in counterintuitive ways, but if you follow the guidelines listed in this section, you will be able to adjust RGB colors to produce the map symbols you want to see on screen.

5

COLOR DECISIONS
FOR MAPPING

CHAPTER 5

COLOR DECISIONS FOR MAPPING

Many factors affect the colors you choose for map symbols. The perceptual structuring of the colors should correspond with the logical structuring in the data. When designing maps, remember that datasets have sequential, diverging, or qualitative arrangements. You can reflect these arrangements—and make your map easier to read—by ensuring that the character and organization of the colors on your map match the logic in your data.

When choosing map colors, you should not be overly concerned about which colors your audience likes. Everyone has an opinion about color aesthetics, and members of your audience will undoubtedly have differing opinions based on their own color preferences. There has been a substantial amount of loosely structured research on color preferences. Regardless of context, it seems that most people like blue and do not like yellow, but that is an overly simplistic guideline for multicolor contexts. People also like maps with many colors, so focus your attention on presenting your data clearly and do not worry about whether you have picked everyone's favorite colors.

Map designers who build effective and efficient maps should consider the following:

▫ color schemes for maps, including sequential, diverging, and qualitative schemes

▫ adjusting color selections for simultaneous contrast, color-blind readers, and black-and-white reproduction

▫ using custom color ramps

Color schemes for maps

When people read your color maps, they use the perceptual dimensions of color, even though you may have specified colors in a mixture system like RGB. Your readers are seeing and thinking about color as "light desaturated blues," "dark saturated oranges," or "dark grays," not as strings of numbers. By using perceptual dimensions in ways that parallel the logical structures in the map's data, you make it easier for your readers to understand the way the information is organized.

Sequential, diverging, and qualitative schemes are used to structure color symbols to correspond with simple data arrangements. More complex data can be mapped by overlaying and combining these three schemes.

Sequential schemes

The most basic guidance for color use on maps is to use lightness to represent ordered data. Ordered data may be numerical data or ranked data. For example, two city populations are ordered whether listed as "50,000 and 650,000 people" or simply as "low and medium."

Generally, darker colors are used to represent higher data values, and lighter colors represent lower values.

In this parcel map, high levels of water use are shown in a dark color. The colors get progressively lighter as usage decreases. This progression is referred to as a sequential color scheme.

Figure 5.1 *Data for water use by parcel demonstrates light-to-dark colors that parallel low-to-high data ordering. In this case, ranked classes rather than numerical data ranges are shown on the map.* Source: ESRI ArcGIS Sample Maps data.

Water Use by Parcel
- High
- |
- Medium
- |
- Low

Sequential color schemes may include hue variation, but they should first and foremost rely on variations in lightness. In the map illustrations, hue changes with lightness. The map showing water use *(figure 5.1)* has symbols that progress from dark blue through medium purple to light pink. The scheme for the accessibility maps that follow *(figures 5.2 and 5.3)* varies from dark blue to light green.

In maps that compare opposing or reciprocal variables, it may be easier to see their relationship if the convention of using darker colors to represent higher values is reversed for one variable. In figure 5.2, it is easier to see the relationship between child mortality and access to markets and infrastructures if the sequential schemes are organized so that low access and high mortality are both represented by dark colors.

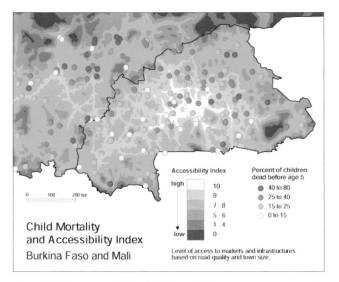

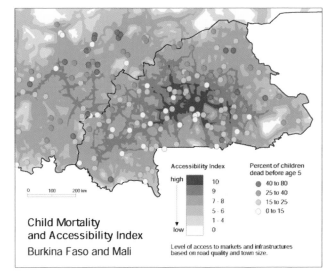

Figure 5.2 *The message of this map is that mortality increases as access to markets and infrastructures decreases. The accessibility lightness sequence is reversed from the standard "dark represents high" convention to emphasize the link between these two variables. High accessibility is represented in light colors while high child mortality is represented with dark colors.* Source: GRID-Arendal, and cartographers Emmanuelle Bournay and Philippe Rekacewicz.

Figure 5.3 *High child mortality and high accessibility are both shown with dark colors. But this strict adherence to the dark-means-high convention obscures the link between low access and high mortality.* Source: GRID-Arendal, and cartographers Emmanuelle Bournay and Philippe Rekacewicz.

Including hue differences with lightness differences while controlling saturation provides contrast between colors, helping your reader to discern them. Figure 5.4 is a selection of example schemes that include hue transitions.

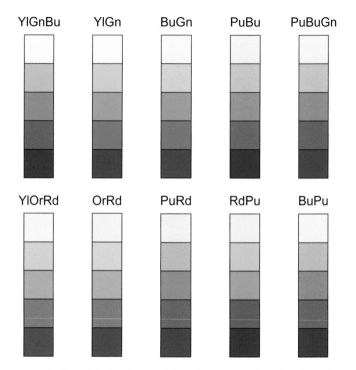

Figure 5.4 *Sequential color schemes included on ColorBrewer.org (see appendix). Each scheme includes a hue transition to complement a lightness sequence. The labels above each scheme indicate the hues used (BuPu indicates blue to purple).*

The color schemes shown here are from *www.ColorBrewer.org,* a Web application that can guide you in selecting map colors. Color-Brewer adapts a selected scheme to the number of classes required for your dataset, then displays it on a sample map. It reports a variety of specifications for each color, including RGB, to allow you to then use the scheme in your mapmaking software. The appendix provides CMYK versions of the ColorBrewer schemes for use on maps that will be printed. RGB conversions of these colors (from Adobe Illustrator) are also listed in the appendix.

Each scheme has been designed to adapt to datasets with three-to-nine classifications. Shown here is the purple-blue-green scheme applied to each of these cases. As the number of classes increases, the more likely it is that adjacent colors in a sequence will be too similar to be discernable in some media, such as a projected image.

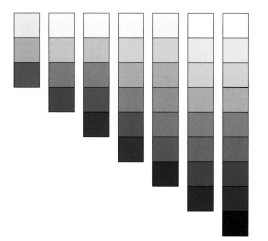

Figure 5.5 *The PuBuGn family of sequential color schemes with an extended hue transition (purple pink, purple, blue, green blue) shown with three-to-nine classes.*

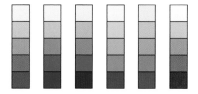

Figure 5.6 *A set of simpler, single-hue, sequential color schemes from ColorBrewer.org.*

Diverging schemes

Careful use of hue and lightness can make maps showing change and difference easy to understand. For example, expected differences from average streamflow may be represented by emphasizing both highs and lows (excessively wet and excessively dry areas). A crucial midpoint in this data range exists where there is minimal difference from usual spring and summer streamflow. A color scheme that emphasizes this midrange with a light color and the extremes with two diverging hues appropriately reflects the nature of this dataset.

Emphasizing the two extremes with darker colors and the average with the lightest color clearly parallels the structure within this dataset. In addition to representing "no difference or no change," this technique can be used to highlight a median, zero, or threshold value.

If the critical value is midway through a class, then that class is represented with the lightest color *(figure 5.8)*. An odd number of classes will result when the critical class is right in the middle of the data range. But, the critical class or break need not be in the middle. The streamflow forecast map has two classes above the average class and three below.

If the critical value is used as a break between classes, then the scheme will have two light colors of different hues straddling that break and no neutral color. The second set of diverging schemes in figure 5.9 shows an even number of classes straddling a critical class break.

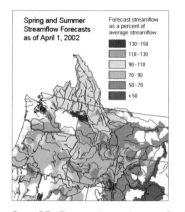

Figure 5.7 *This map showing the streamflow forecasts of drainage basins in the northwestern United States uses a diverging color scheme with two hues (brown and blue green) and a light-gray midrange class for near-average streamflow amounts.* Source: Adapted from maps distributed by National Water and Climate Center (Portland, Oregon), Natural Resources Conservation Service, USDA; www.wcc.nrcs.usda.gov

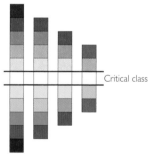

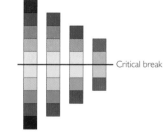

Figure 5.8 *Diverging purple-green color schemes with odd numbers of classes. Each scheme is symmetrically organized around a middle light class that corresponds to a critical range in mapped data.*

Figure 5.9 *Diverging purple-green color schemes with even numbers of classes. Each scheme is symmetrically organized around a middle class break that corresponds to a critical data value.*

Spectral or rainbow schemes are popular in scientific visualizations and news media graphics like daily weather maps. The full sequence is familiar: dark red, red orange, orange yellow, yellow, yellow green, green blue, and dark blue. Unfortunately, this scheme is often misused as a sequential scheme rather than as a diverging scheme. Using it to display a sequential dataset, such as temperature, places unintended emphasis on arbitrary midrange values by representing them with the lightest and brightest color, yellow. The dark red and dark blue endpoints of the scheme should mark the extremes in the data and light yellow should emphasize a meaningful midrange, such as zero change in population or average streamflow.

The most informative use of a spectral scheme is as a diverging scheme. People like the multihue character of the scheme, and the variety of hues helps distinguish symbol categories. Structuring the lightness sequence to parallel the characteristics of the data produces an informative map *(figure 5.10)*.

A dataset can often be examined as both sequential and diverging. Regarding data as high-to-low and as deviations from an average may both be equally meaningful. Diverging and sequential are conceptualizations, rather than absolute properties, of data.

Figure 5.11 shows the same streamflow forecast data using a sequential scheme that emphasizes the forecast percentages in general, rather than how they differ from normal. An advantage of this rendition is that it could be adapted to print in black and white. This sequential scheme retains some emphasis on the middle, near-average class (90 to 110 percent flow) by using a higher saturation color than others in the scheme. Thus, the scheme includes a hue transition (blue-green-yellow), sequential lightness (dark-medium-light), and diverging saturation (middle-high-low saturation).

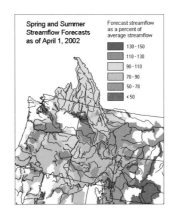

Figure 5.10 *Streamflow forecasts shown with a diverging spectral color scheme. The lightest color, yellow, is used to represent the near-average class.* Source: Adapted from USDA map source.

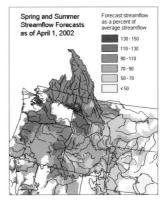

Figure 5.11 *Streamflow forecasts shown with a sequential (light to dark) scheme, rather than a diverging scheme. Notice that the near-average midrange is subtly emphasized using higher saturation.* Source: Adapted from USDA map source.

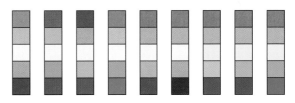

Figure 5.12 *The selection of diverging schemes available in ColorBrewer.org. Most offer two hues in lightness ramps joined at a shared light color. The last three include a wider range of hues.*

Qualitative schemes

Qualitative schemes represent different kinds of map features or categories that are not ordered. Categorical differences in data are usually represented with differences in hue. For example, types of government spending, such as military, education, and healthcare, are categories that could be shown with unique hues like red, green, and blue. A well-designed scheme will not suggest that data is ordered or that one category is erroneously more important than another. To maintain this sense of similar importance, ensure that hues in a qualitative color scheme maintain similar contrast with the background of the map by controlling the lightness and saturation of each color. For example, yellow will not be as visible on a white background as red, green, or blue, so the class you assign to yellow may be perceived as having a different importance than the others.

Maps with small intermixed polygons and more than about five qualitative categories are difficult to symbolize. You will need to make use of all of the perceptual dimensions of color, varying lightness, and saturation regardless of the implied magnitude differences these color differences will suggest. Make these adjustments as intelligently as you are able given the map topic.

If logical relationships exist between categories, echoing those relationships with related colors within the qualitative scheme improves the map. For example, use light and dark red for two types of built-up areas and three different greens for three types of forest on a land-cover map. The groupings of color allow the map to be read for a general overview as well as category by category.

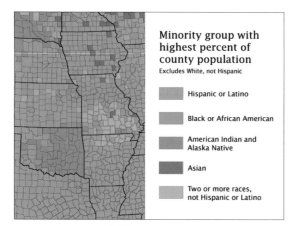

Figure 5.13 *Comparison of the prevalent minority groups by county is an example of qualitative data. The color scheme maintains fairly constant lightness and saturation, varying only hue among the five classes. The minimal contrast in this scheme diminishes the perception of groupings in the data.* Source: Map adapted from Brewer and Suchan. 2001. *Mapping Census 2000: The Geography of U.S. Diversity.* U.S. Census Bureau.

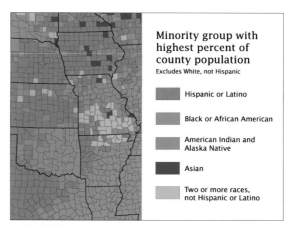

Figure 5.14 *The same map shown with more lightness and saturation contrast between colors. Overall, the colors are not markedly light or dark, but the addition of some lightness contrast assists the reader in identifying county classes and seeing groups. Categories with fewer classes have more contrast: "Asian" is darker than the others and "Two or more races" is more saturated than the others.* Source: Map adapted from Brewer and Suchan. 2001. *Mapping Census 2000: The Geography of U.S. Diversity.* U.S. Census Bureau.

When you have small areas on a map, choose high-saturation colors to emphasize them and to make sure the colors can be identified. The small wetland and crop cover types are shown in lighter and more saturated colors than those of larger areas to make them more visible *(figure 5.15)*.

One way to choose colors for qualitative categories is to use hues already associated with each category, for example, red for tomatoes and yellow for corn. But do not become overly concerned about figuring out the perfect associations between map categories and hues. More often, there are no obvious hue relationships for the abstract topics common to thematic maps. Thin associations will be understood by only some of your map readers, so are not worth laboring over. Use hue as an abstract symbol and focus your energies on making the colors easy to differentiate on the map.

A reason not to be adamant about hue associations is they may not be understood by everyone who reads the map. Sometimes a seemingly obvious hue association for you may not be so obvious to other groups. For example, red may seem just right for a hotspot on your map, but red may indicate loss to a more accounting-oriented audience ("in the red"), producing an unintended association. Green might represent money to a U.S. mapmaker, but that link may not be meaningful to an audience whose paper currency is not green.

Figure 5.15 *Qualitative land-cover map with varied lightness and saturation for map symbols. Reds and greens are used to group related classes. High-saturation purple, yellow, and cyan are used to emphasize small land-cover classes.* Source: National Land Cover Dataset, USGS, 1992.

Be alert for color associations that may be offensive. For example, you may need to exercise caution if a particular hue is associated with a controversial political party in the region mapped. Also, take care with some literal uses of color, such as black for people who are African American, yellow for Asians, and red for American Indians. The superficial and exaggerated emphasis on skin color associations for groups is likely to offend your readers. Use a purposely abstract set of hues for mapping race groups instead. Work to develop a set of easily distinguished hues to symbolize your data, but also be astute about unintended meanings individual hues may have for the topic mapped.

The qualitative schemes in ColorBrewer range from three to twelve colors and vary in lightness. The second-to-last scheme (*figure 5.17*) includes lightness pairs within hues to use for qualitative data that includes related categories.

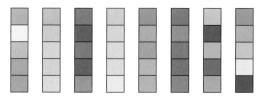

Figure 5.17 *A series of five-class qualitative schemes from ColorBrewer.org. Families of schemes with up to twelve classes are offered by the online color tool.*

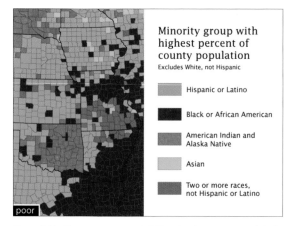

Figure 5.16 *Do not use stereotypical skin colors for race map symbols. Overemphasis on superficial characteristics detracts from the map message.* Source: Map adapted from Brewer and Suchan. 2001. U.S. Census Bureau.

Binary schemes are used to symbolize the simple case of qualitative data with just two classes. You may use hue, lightness, and saturation to differentiate two map classes. Put more visual emphasis on one class if it is more important to the map. The examples below emphasize areas that can be seen from a particular viewpoint versus areas that are hidden from view *(figure 5.18),* and built-up areas versus other land cover *(figure 5.19).*

Observation point

Figure 5.19 *A binary scheme of dark red and light green contrasts built-up land with other land-cover types (crop land, forest, barren land, and so forth) in the area surrounding Charleston, West Virginia.* Source: USGS.

Figure 5.18 *A viewshed for the west half of Crater Lake, Oregon, calculated for a point slightly below the crater's rim. Areas that can be seen from the observation point are emphasized with blue and areas that remain hidden from this location are beige. Shaded relief accents the landforms in the area. The south half of cone-shaped Wizard Island and the lake surface behind the island are not visible.* Source: USGS Crater Lake Data Clearinghouse, *craterlake.wr.usgs.gov/intro.html*

Adjusting color selections

There are varied reasons to fine-tune color selections to accommodate map readers. The apparent lightness, hue, and saturation of a color can be altered by surrounding colors, often producing unexpected color changes. It is important to adjust your design to anticipate these changes. Additionally, color maps sometimes need to be read without their intended hues. If your map readers may be color-blind or your map is photocopied in black and white for inexpensive distribution, lightness differences are the key to retaining as much map information as possible. GIS color ramps can be improved by using perceptual color design skills and customizing the match between ramp characteristics and data values.

Unexpected color changes

Color appearance is affected by context. Small colored objects on a map are more difficult to identify than large colored areas: you will be able to distinguish fewer colors when the map includes small point symbols or thin lines. Different surroundings can also change the appearance of a color.

The contrast between a patch of color and its surrounding color enhances the difference between the two colors in a process called simultaneous contrast. For example, a color of medium lightness will look darker on a light background and lighter on a dark background. Hue changes are toward opponent complements: red–green and yellow–blue. A gray will look greenish on a red background. A green will look lighter and yellower on a dark blue background. The saturation component of color is particularly susceptible to being changed by simultaneous contrast.

A series of examples based on simple diagrams and two maps *(figures 5.20, 5.21, and 5.22)* demonstrate the relative nature of color and how it changes with differences in background. The maps and diagram all share the same set of colors: a sequential scheme that ranges from light yellow to dark blue. The map scheme has eight steps, which is a lot, so problems with simultaneous contrast effects are expected.

Observe how colors are affected by their surroundings, and test the map schemes by seeing whether you can match the identified colors to the legend despite their varied backgrounds. These sorts of changes also affect color comparisons between regions on a map and between maps in a series. Can you decide which counties represent the same data range on both maps?

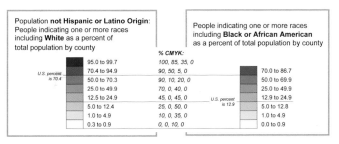

Figure 5.20 *Two map legends that share the same color scheme. The legend on the left is for the map segment shown in figure 5.21, and the legend on the right is for the map shown in figure 5.22. These two maps (right) are used to demonstrate simultaneous contrast. Colors in CMYK percentages are listed between the two legends. The two maps share the same colors and similar data classifications.*

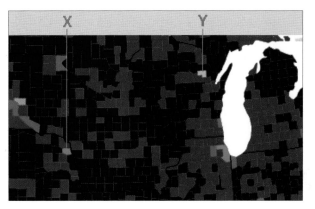

Figure 5.21 *Two isolated colors have dark surroundings; they are identified as X and Y above the map. Which legend color do they match in figure 5.20? Which of the four colors in the following map (A to D in figure 5.22) do X and Y match? The map shows percent of total population who indicated non-Hispanic White race/ethnicity for Census 2000.* Source: Brewer and Suchan. 2001. U.S. Census Bureau.

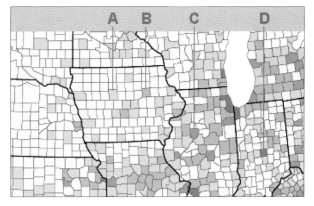

Figure 5.22 *These counties have light surroundings, which make them look darker. Can you match them with the legends in figure 5.20, and with X and Y in the above map (figure 5.21)? (See answer in figure 5.24.) The map shows percent of total population who indicated Black or African American race for Census 2000.* Source: Brewer and Suchan. 2001. U.S. Census Bureau.

The same colors are shown below in a simpler arrangement, with and without homogeneous colorful surroundings, so you can see which ones are made to look similar and how matched colors look different depending on surround effects.

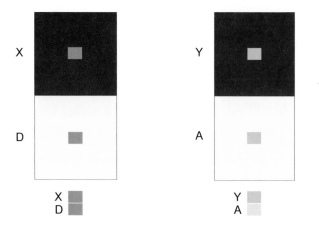

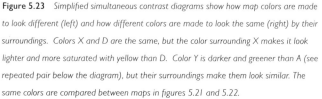

Figure 5.23 *Simplified simultaneous contrast diagrams show how map colors are made to look different (left) and how different colors are made to look the same (right) by their surroundings. Colors X and D are the same, but the color surrounding X makes it look lighter and more saturated with yellow than D. Color Y is darker and greener than A (see repeated pair below the diagram), but their surroundings make them look similar. The same colors are compared between maps in figures 5.21 and 5.22.*

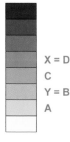

X = D
C
Y = B
A

Figure 5.24 *Color X matches D, though its dark surroundings on the first map make it look like C, a lighter color. Color Y matches B, though simultaneous contrast makes it look like A on the map (compare figures 5.21 and 5.22).*

Simultaneous contrast makes a mess of the map below. The gray tracks shift toward blue because they are surrounded by yellow. The blue ferry routes, surrounded by a darker blue, shift lighter and more yellow to look grayish. The yellow roads shift lighter and bluer, so they look like a very light gray. In the end, ferry routes look like tracks in the legend, tracks look like ferry routes, and roads don't seem to match any legend symbol. The different surroundings for symbols on the map and in the legend make the legend almost worthless.

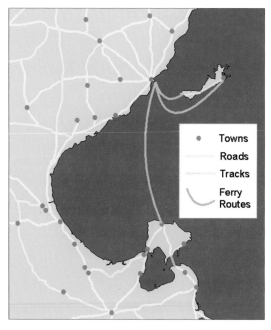

Figure 5.25 *Varying backgrounds shift the appearance of desaturated line colors.*

Source: ESRI, Digital Chart of the World.

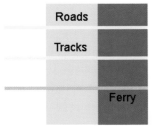

Figure 5.26 *A schematic diagram shows the changes in line colors on different backgrounds for the map in figure 5.25. From left to right, the yellow line looks grayer and then lighter on the two backgrounds. The gray line looks bluer, then lighter. The blue line looks bluer, then grayer.*

There is little that can be done to prevent contrast effects from occurring because the distribution of data controls the positions of the symbols. The best approach is to look carefully at the colors you are planning for a map. Do not select colors by comparing them only in the legend where they are seen in one order on a uniform background. Look at your selection of colors in the final map pattern and in the final format (as a projected slide or overhead transparency, color inkjet print, lithographic proof, color photocopy, and so forth). Be sure you can identify examples of each color symbol and can tell colors apart throughout the map. Look for isolated areas where one color is surrounded by contrasting colors; these situations truly test a scheme.

A critical eye and good color contrast counteract simultaneous contrast on maps.

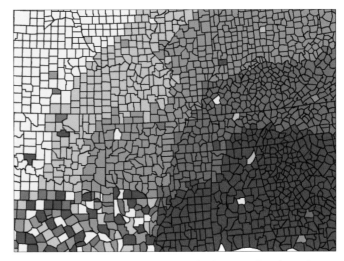

Figure 5.27 *The diagnostic map display used in ColorBrewer.org, shown here with a five-color sequential scheme. Each color is isolated in a band of each of the other colors. If the scheme is well designed, you should be able to easily find the four other colors within each band.* Source: www.ColorBrewer.org

Colors for the color blind

Approximately 4 percent of the population have some degree of color vision impairment (approximately 8 percent of men and less than 1 percent of women). Color-blind people can see lightness differences and a fairly wide range of hue differences.

Red-green color blindness is the most common impairment, but it includes confusions between other hue combinations as well, such as magenta and cyan. The severity of color blindness varies from person to person. Desaturated colors, like rust and olive, are more difficult for people with milder color vision impairments to distinguish than saturated red and green.

Expected color confusions can be rigorously modeled in color order systems, but these results are difficult to apply without color measurement equipment. Guidelines specified by color names, rather than color measurements, eliminate a wider range of color combinations than necessary, but they are useful in the context of designing maps. The following pairs of hues are not confused by people with the most common types of color vision impairments:

- red and blue
- red and purple
- orange and blue
- orange and purple
- brown and blue
- brown and purple
- yellow and blue
- yellow and purple
- yellow and gray
- blue and gray

These ten color pairs are from a total of thirty-six pairs of basic color names, so many other hue pairs are confusing. For example, any combination of red, orange, brown, yellow, and green is potentially confusing if the colors have similar lightness. Any combination of magenta, gray, and cyan is also likely to be indistinguishable to color-blind people.

These guidelines are intended to be used to design a map by simply naming colors, rather than specifying positions in a color space. For example, choosing a red-blue pairing includes any colors that you, or your client, would call red and blue, without regard to what sort of red and blue is chosen (yellowish red or bluish red, for example).

The map in figure 5.28 includes many hue combinations that will be difficult for color-blind map readers to discern. The magenta ferry routes may be hard to see against the cyan water. The orange and red roads may be hard to tell apart and hard to see on the green land. The brown town symbols may also be hard to see against green. Poor legibility for any reader is exaggerated by the lack of lightness contrast between symbols and their backgrounds, but these symbols will be nearly invisible for color-blind people.

Guidance more specific than general color name pairs can be derived from a variation on the color circle you were using in the previous chapter *(figure 5.29)*.

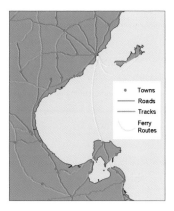

Figure 5.28 *A portion of a map of Tunisia demonstrates poor choices in colors because many will look the same to color-blind map readers. The road, track, and town colors will all look like each other and will be difficult to distinguish from the background. The pink for hypothetical ferry routes is likely to look the same as the surrounding cyan water.* Source: ESRI, Digital Chart of the World.

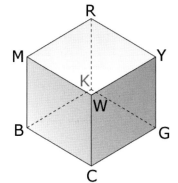

Figure 5.29 *The CMY color cube with the white corner centered. Hues are arranged in spectral order around the edge of the cube in this view. Figure 5.30 shows a sampling of colors across the visible faces of this cube.*

Flattening and fleshing out the cube to show colors on its three visible surfaces presents a hue circle with a white center, shown in figures 5.30, 5.31. and 5.32. I have built an approximate set of confusion zones across this color diagram based on color-blind confusion lines through a more technical color space called CIE xyY, which is difficult to use for map color specification.

Colors with similar lightness that are in the same zone or adjacent zones of figure 5.30 are likely to look the same to a color-blind person. Colors chosen two or more zones apart will be easier to discern. Figure 5.31 shows how to use the diagram. Colors from the purple-blue zone will not look the same as colors from the red-yellow-green area. Colors that are different in lightness within either zone—or adjacent zones—will be visibly different as well.

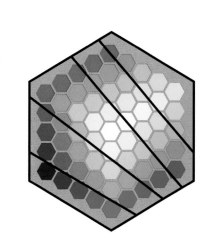

Figure 5.30 *The black lines across the diagram outline color-blind confusion zones. The underlying arrangement of colors is built from all combinations of 0, 10, 30, 60, and 100 percent of cyan, magenta, and yellow inks (see page 199 in the appendix for percentages for individual colors and conversions to RGB).*

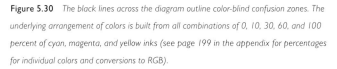

131

The selection of six colors, shown with black outlines in figure 5.31, forms a workable diverging scheme for both color-blind readers and readers with normal color vision. The red-orange-yellow colors differ in lightness, so they can be used even though they are from the same zone. Likewise, the blue-purple pair includes a lightness difference. Colors with similar lightness (orange and blue, for example) are from separated zones. This scheme is a good, colorful replacement for a spectral scheme because it skips over the greens.

All types of color schemes—sequential, diverging, qualitative—can be adjusted so they can be read by the color blind. Almost any well-designed sequential scheme that has good lightness contrast between colors will accommodate color-blind readers. Beyond visualizing the logical structure of the data, accommodating map readers with color vision impairments is another reason to use systematic lightness steps when representing ordered data.

A basic strategy for designing diverging schemes is to choose pairs of hues from the list on page 130 and build a lightness sequence within each hue. Saturated yellow does not include dark colors, so it offers few lightness steps. The small legends shown in figure 5.32 are good examples of hue pairs for diverging schemes. Diverging spectral schemes that skip greens are also good for color-blind readers.

Designing qualitative schemes for people with color vision impairments is difficult because these schemes require many different hues. Careful use of lightness differences and separation among confusion zones can produce a qualitative scheme that accommodates most color-blind readers. For example, orange and blue in a scheme can have the same lightness and still be legible to color-blind readers. But if green is also used, it should be either lighter or darker than the orange.

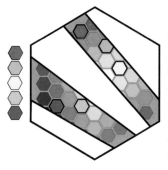

Figure 5.31 *To use the color-blind confusion diagram, choose colors separated by at least one zone, or colors that are different in lightness. The selected red-orange-yellow colors (with black outlines) are from the same zone but differ in lightness. Selected colors with similar lightness are two or more zones apart. A diverging scheme with six colors was constructed using the diagram (arranged at left).*

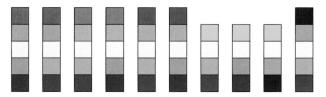

Figure 5.32 *Diverging schemes built from hue pairs, or gray paired with a hue, that are distinguishable to color-blind map readers.*

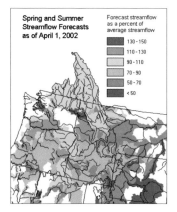

Figure 5.33 *The streamflow map with a near-spectral diverging scheme that skips over greens to accommodate color-blind readers.* Source: Adapted from USDA map.

Finding a color-blind person to look at your maps is helpful. With 8 percent of men affected, a group of twelve fellows is likely to include one who is color blind. You will have a much harder time finding a color-blind helper in a group of women given the genetics of congenital color blindness. Another way to evaluate colors is to use the online utilities at *www.vischeck.com* to simulate what color-blind people will see when looking at your map.

The severity of color blindness varies widely, and there are a few different types of color blindness. Thus, you do not want to lull yourself into a false sense that you have designed for all color-blind people when you have accommodated a single color-blind person. Having someone around to vet the schemes is helpful, but it is not complete assurance that the map will work for all color-blind readers.

Many people with mild color blindness do not know they are color blind. You may find yourself making adjustments to seemingly good schemes to accommodate a difficult colleague, when in reality you are accommodating his vision impairments. Watch which colors the person is objecting to and you may be able to make an amateur diagnosis (though perhaps you will keep it to yourself). You can be proud of your ability to make color maps that the colleague and a wider audience of color-blind map users are able to read using the guidance presented here.

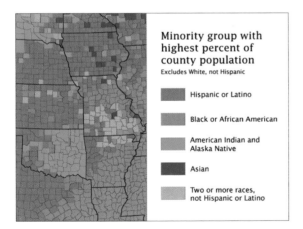

Figure 5.34 *This qualitative scheme does not accommodate color-blind map readers. The red, orange, and green are likely to look the same.* Source: Map adapted from Brewer and Suchan. 2001. U.S. Census Bureau.

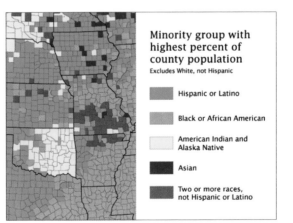

Figure 5.35 *Here, the qualitative scheme has been redesigned with greater lightness differences and careful hue choices. Yellow-green-brown would normally be confusing, but here they differ in lightness, as do the blue and purple. Color pairs with similar lightness—purple-green and blue-brown (brown is a desaturated orange)—are well separated on the confusion diagram.* Source: Map adapted from Brewer and Suchan. 2001. U.S. Census Bureau.

Photocopying color maps

The key to making a color map that can be printed or photo-copied in black and white is to ensure large differences in light-ness between colors. You can think of the photocopier or laser printer as your ultimate color-blind reader. If your map is still readable after it has been photocopied, then the color blind will also be able to read it.

Copiers can reliably produce only about four gray levels: white, light gray, dark gray, and black. This is a very limiting constraint, so you may decide to redesign a color map explicitly for black-and-white reproduction rather than trying to have one map suit both purposes. Judicious use of patterns that will copy or print in black and white may improve the map for reproduction.

Copiers do better with grays that are constructed from distinctly visible dot patterns. Modern laser printers are able to print grays with such fine dots (for example, 600 dpi) that photocopiers are not able to resolve them to reproduce these grays. If you are able to control print resolution using dots-per-inch (dpi) or lines-per-inch (lpi) settings, choose a coarser resolution so the dots that build up gray tones are visible. Coarse dots will give you a more reliable reproduction of the map when it is copied. For example, 300 dpi printing or 70 lpi screens can be accurately reproduced by most photocopiers.

If you are preparing maps to be reproduced in-house, you will achieve your best results by testing your maps on the copier that will be used to reproduce them. Copier quality varies widely and some hues copy darker than others. Make sure your color choices are suitable before you make many copies. Adjust colors as neces-sary until the photocopied map is clearly readable. This is labori-ous but necessary if you do not want to risk distributing hundreds of copies of indecipherable maps. Reproduction by color photo-copying can be just as uncooperative, so prepare with a few trial runs. Light colors are most likely to suffer with either black-and-white or color copying.

A selection of maps from previous pages have been converted to grayscale to see how well their color schemes fare in monochrome. These digital files provide better quality than photocopying, but can give you a sense of which schemes will hold up without hue.

Sequential schemes do fairly well, though the contrast between grays is much less than the contrast between the hue/lightness/saturation combinations seen previously. Compare figures 5.36 and 5.37 to color maps earlier in this chapter.

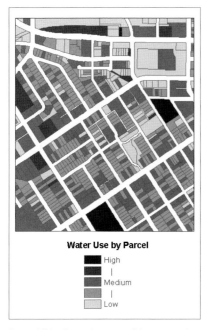

Water Use by Parcel

- High
- |
- Medium
- |
- Low

Figure 5.36 *Grayscale version of the sequential water use map (figure 5.1). The light-to-dark sequence holds up without its hues.* Source: ESRI ArcGIS Sample Maps data.

Accessibility Index		Percent of children dead before age 5
high	10	40 to 80
	9	25 to 40
	7 - 8	15 to 25
	5 - 6	0 to 15
	1 - 4	
low	0	

0   100   200 km

Child Mortality and Accessibility Index
Burkina Faso and Mali

Level of access to markets and infrastructures based on road quality and town size.

Figure 5.37 *Grayscale version of the sequential mortality and accessibility map (figure 5.2). Though contrast is reduced, the match in the lightness ranges is enhanced: light low-mortality dots often fall on light high-access areas.* Source: GRID-Arendal, and cartographers Emmanuelle Bournay and Philippe Rekacewicz.

Diverging schemes do poorly in grayscale. The dark ends of the scheme lose the contrast created by different hues. When reproduced in black and white, maps relying on these schemes are reduced to emphasizing deviation from the middle light class without stating which classes are above or below the middle (*figure 5.38*).

Binary qualitative schemes fare well in monochrome since only two classes, light and dark, are needed.

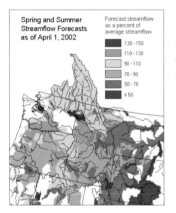

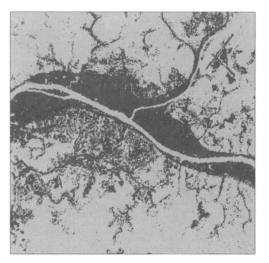

Figure 5.38 *The diverging streamflow map (figures 5.10 and 5.11) no longer delivers its intended message because you cannot tell the difference between highs and lows in the grayscale version.*
Source: Adapted from USDA map.

Figure 5.39 *The binary scheme emphasizing built-up land cover works fine in grayscale (color version in figure 5.19).* Source: USGS.

Qualitative schemes do poorly when limited to grayscale because the inevitable light-to-dark ordering of monochrome symbols suggests a category ordering. The land-cover map with many different hues ends up with unrelated classes in very similar grays when it is converted to grayscale.

Land-cover classes may be more carefully collapsed to produce a somewhat useful grayscale map. Figure 5.41 shows five classes.

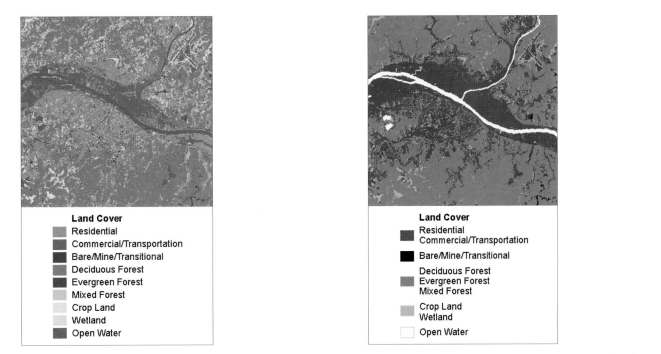

Land Cover
- Residential
- Commercial/Transportation
- Bare/Mine/Transitional
- Deciduous Forest
- Evergreen Forest
- Mixed Forest
- Crop Land
- Wetland
- Open Water

Land Cover
- Residential
  Commercial/Transportation
- Bare/Mine/Transitional
- Deciduous Forest
  Evergreen Forest
  Mixed Forest
- Crop Land
  Wetland
- Open Water

Figure 5.40 *The qualitative land-cover map is no longer useful in grayscale because unrelated classes look the same after a direct conversion (color version in figure 5.15).* Source: USGS.

Figure 5.41 *Redesign of the qualitative land-cover map in grayscale produces a more limited but more effective display (color version in figure 5.15).* Source: USGS.

Custom color ramps

The example maps earlier in this chapter present sets of colors with discrete classes, such as a five-class sequence ranging from yellow to red. An alternative strategy for designing a color scheme using GIS is to select endpoint colors and automatically "ramp" between them. The ramp may be presented as a continuous color gradation on a map or may be split into a series of discrete colors for map classes.

Because of problems with routes through color space selected by the software, ramping does not always produce quite what you intend. For example, a ramp from yellow to blue cuts through color space to desaturated colors midway through the ramp. This is a logical straight-line route through the color space, but may not produce an appealing map.

If you would like more saturated colors in the ramp, you need to choose a route that arcs outward rather than going straight through the middle of the color space. You can approximate this by specifying interim colors as well as endpoints. A straight yellow-to-blue ramp includes some muddy midrange colors, but specifying a midrange green, as shown in the second map in figure 5.42, improves the scheme. Likewise, including an orange midway between yellow and red will produce a ramp with colors that are easier to tell apart and are more pleasing in appearance *(figure 5.43)*.

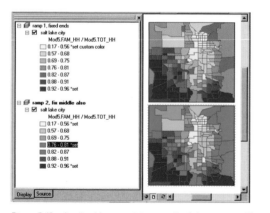

Figure 5.42 *A yellow-blue ramp is improved by fixing a green midpoint before ramping in ArcMap. Yellow and blue are set for the end colors in both ramps, but the second ramp also has a green set at the middle (highlighted in the legend). Ramp 2 runs from yellow to green and from green to blue, rather than directly from yellow to blue. The map shows Census 2000 data for Salt Lake City census tracts: family households as a percent of total households.* Source: U.S. Census Bureau.

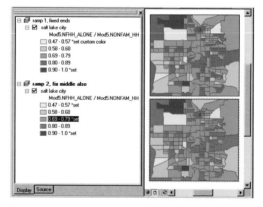

Figure 5.43 *A yellow-red ramp is improved more subtly with this technique. A saturated orange midpoint was set in the second ramp. It is highlighted in the legend (in ArcMap, press the Control key with the midrange color selected while you create the ramp). The map shows Census 2000 data for Salt Lake City census tracts: households with one person as a percent of all nonfamily households.* Source: U.S. Census Bureau.

The strategy of fixing midrange colors for ramps also allows you to design flexible diverging schemes. Custom ramping gives you a flexible number of classes to make a logical lightness and hue structure. In figure 5.44, white symbolizes the 0.91 to 1.10 class, straddling 1.00 (equal number of people sixty-five and over in family households and nonfamily households). Preset ramps do not provide enough control to symbolize classes that are asymmetrically arranged above and below a critical midrange data class.

Ramping is useful for reducing the tedium of explicitly specifying colors for numerous classes. If you have so many classes, however, that typing color specifications is time-consuming, you probably have too many color classes to be differentiated by the map reader. Consider your reader and your design goals carefully before you concoct a map with more than about ten classes.

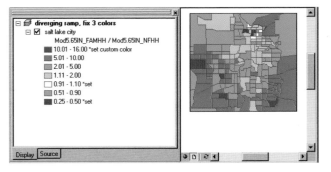

Figure 5.44 *Dark red, white, and dark blue are used to make this asymmetrical diverging scheme. The white class represents nearly equal numbers of elderly in two household situations. The map shows Census 2000 data for Salt Lake City census tracts: the ratio of people sixty-five and older in family and nonfamily households.* Source: U.S. Census Bureau.

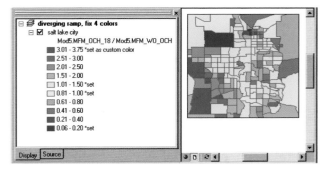

Figure 5.45 *A diverging ramp without a middle light class. Four colors were set (light and dark for each hue) before creating the ramp. The map shows Census 2000 data for Salt Lake City census tracts: the ratio of married family households with their own children under eighteen at home to married families without their own children at home.* Source: U.S. Census Bureau.

Figure 5.46 shows forty classes; thirty-six between white and dark red. This approach produces an essentially continuous gradation, called an n-class or unclassed choropleth map. Unclassed choropleth mapping is useful in some contexts, but not when your readers need to identify data ranges for individual areas on the map.

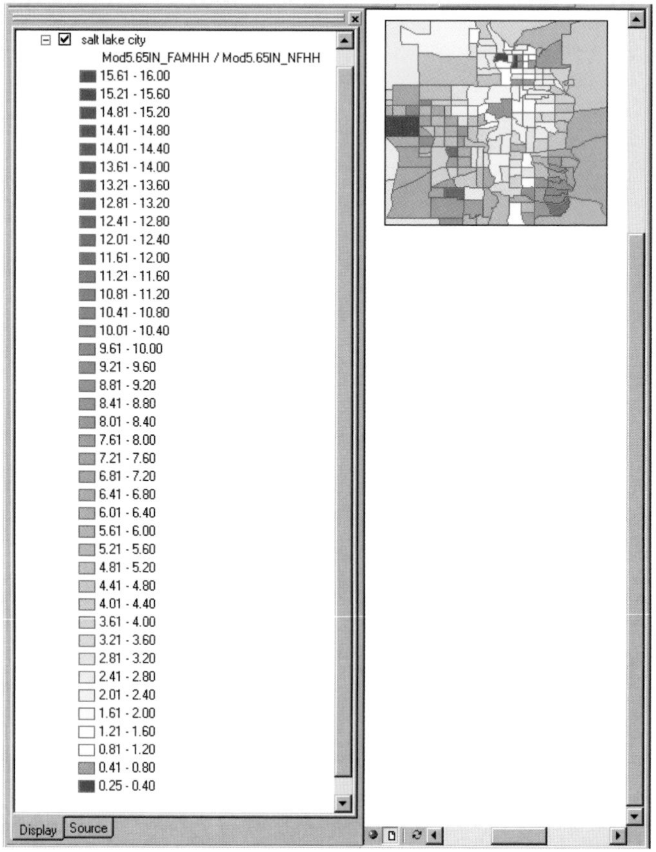

Figure 5.46 *An n-class choropleth map. Be cautious about creating so many classes that readers cannot identify data ranges.* Source: U.S. Census Bureau.

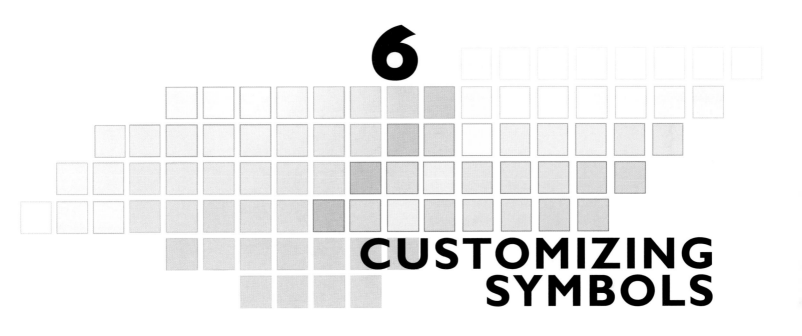

6

CUSTOMIZING SYMBOLS

CHAPTER 6

CUSTOMIZING SYMBOLS

The subtlest of details can determine how map data is read and interpreted. The shape of a marker, the width of a line, the arrangement of a pattern—each conveys specific information. Map designers pay careful attention to the match between visual variables (such as shape and size) and the characteristics of their data when creating customized symbols.

There are two basic groups of visual variables that structure symbol design. Lightness, size, pattern separation, and saturation are used when symbolizing ordered data. Hue, shape, and pattern arrangement categorize map features into qualitative classes.

The combination of visual variables and feature types (point, line, and area) forms a general framework that you can use to design customized symbols for representing all kinds of map data. Using multiple visual variables together within symbols, such as hue with shape, produces an almost endless variety of map symbols. There are few excuses for using a default symbol that does not match well with the data it represents because you can make almost any map symbol you can envision.

In addition to using color symbols, map designers work with the following:

- point symbols by varying symbol size, shape, and angle

- line and area symbols by varying line size and pattern and by constructing area patterns

- visual variables for point, line, and area symbols that represent ordered data and qualitative data

Point symbols

Point symbols are not limited to representing data linked to point features on a map. Whether a feature is a point or an area depends on the map scale. A city may be a point at a small scale and an area at a larger scale. Likewise, an address may be a point on a map of a city but an area on a larger-scale parcel map. Thus, you want to think of "point symbols" as flexibly applying to both point and area locations.

In addition to hue and lightness, the main visual variables used to create point symbols are size, shape, and angle. Pictograms used as point symbols are a special case of shape differences.

Point symbol size

Symbol size is used to represent data values at point locations or for areas. Larger symbols represent higher data values. The simple examples below show water use and illumination strength provided by street lighting. Symbols are centered in parcel areas to represent water use for whole parcels in figure 6.1. In figure 6.2, they are located at lamppost point locations.

Figure 6.1 *The visual variable size is used to represent data values for areas. Water use by parcel is mapped with larger symbols representing parcels with higher water use.* Source: ESRI ArcGIS Sample Maps data.

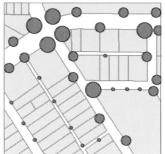

Figure 6.2 *The visual variable size is used to represent data values at point locations. Here, hypothetical illumination amounts from individual street lamps are mapped with larger symbols for greater illumination.* Source: ESRI ArcGIS Sample Maps data.

Point symbols that vary in size may be used to represent either total amounts or derived values. For example, sizes may represent the number of deaths or the death rate from a disease. It is more common to use point symbols for totals *(figure 6.3)*; choropleth maps are the main method of representing rates and other derived values *(figure 6.4)*.

When you use point symbols to represent area data, they are usually located in the middle of each map polygon. The symbol may be smaller or larger than the polygon area, depending on the data value it represents. In contrast, when color alone is used, the area of the symbol on the map is determined by the size of the map polygon. High data values for small polygons, such as high-population urban areas, may have a much reduced visual impact on a choropleth map, and the data may be represented better using symbol size.

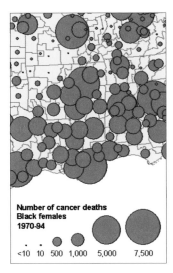

Figure 6.3 *Total number of deaths among Black females from 1970 to 1994 in the lower Mississippi River region by state economic area (SEA). This data is represented with proportioned point symbols. Symbol areas are proportioned to data values.* Source: www.cancer.gov/atlasplus

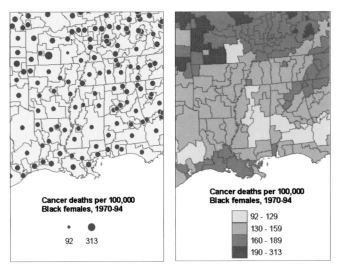

Figure 6.4 *Age-adjusted mortality rates for Black females over the same time period shown in figure 6.3. Data is mapped with proportioned point symbols (left) and area symbols (right). The range in this data is narrow, so symbols proportioned by area do not vary much in size. This rate data may also be represented using lightness and hue on a choropleth map. Classing used in the choropleth map accentuates spatial pattern in this data.* Source: www.cancer.gov/atlasplus

Mortality data is shown below with point and area symbols. The point symbols are filled with the same colors used for the area symbols to emphasize the link between the two representations.

Both proportioned and graduated symbols use differences in size to represent data. Proportioning symbol sizes to map data gives you a fairly exact understanding of differences in magnitude between data values. A city with twice the population of another will have a proportioned symbol twice as tall or with twice the area. If the symbols are graduated, the symbol for the larger city will simply be a bit larger, but it will not attempt to show how much larger the city is. Graduating (or range grading) symbols is a less exacting method that ranks values rather than representing data amounts. Notice the marked difference in the size range between the symbols representing ten and five thousand deaths in the maps in figure 6.6. Proportioned symbols represent data magnitudes and graduated symbols use size to represent data order.

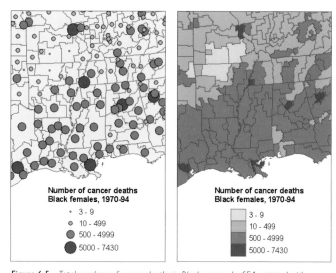

Figure 6.5 *Total numbers of cancer deaths in Black women by SEA mapped with graduated point symbols (left) and area symbols (right). The same graduated color set is used to emphasize the identical classification of the data in these two representations.*

Source: *www.cancer.gov/atlasplus*

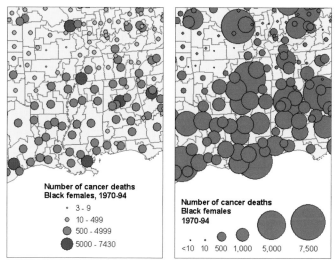

Figure 6.6 *Graduated (left) and proportioned (right) point symbols. Note the smaller range in sizes among the graduated symbols (left). Each of these symbol sizes represents a data range, rather than being proportioned to specific data values.* Source: *www.cancer. gov/atlasplus*

The examples shown previously in this section have used circle areas to represent data. Lengths of linear point symbols and areas of shapes other than circles are also used for constructing map symbols. Proportioned squares and bars are shown in figures 6.7 and 6.8.

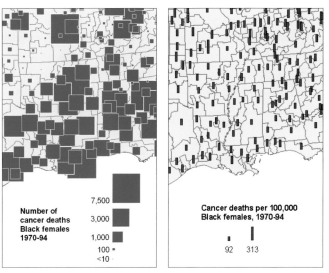

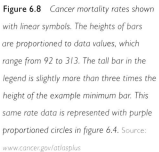

Figure 6.7 *Number of cancer deaths shown with proportioned squares. The areas of squares are proportioned to the data values. Note that white outlines may be used to separate symbols with dark fills. These same data totals are represented with saturated red circles in figure 6.6.* Source: *www.cancer.gov/atlasplus*

Figure 6.8 *Cancer mortality rates shown with linear symbols. The heights of bars are proportioned to data values, which range from 92 to 313. The tall bar in the legend is slightly more than three times the height of the example minimum bar. This same rate data is represented with purple proportioned circles in figure 6.4.* Source: *www.cancer.gov/atlasplus*

Point symbol shapes

Shape is another visual variable readily used with point symbols. Shapes may be simple geometric forms (such as squares, circles, and triangles) or more complex. Shape is used to represent qualitative differences in data values. A simple example map uses a star for a local information office, a circle for the post office, and a triangle for the library *(figure 6.9)*.

Symbol shapes can also be used to build intricate codes that vary in form to represent qualitative data. Pictograms (or icons or glyphs) make more elaborate use of shape for data representation. Common sets of pictograms are offered in the styles and the fonts that are installed with ArcMap. The same simple map is shown with familiar and readily distinguished pictograms *(figure 6.10)*.

There are few guidelines for designing or selecting shapes for map symbols. If there are symbol conventions for a map topic, such as picnicking and camping pictograms *(figure 6.11)*, it makes sense to use them. Do not expect a reader, however, to be clairvoyant in their ability to interpret your pictograms. Using many small and intricate diagrams all over your map will stymie even the most diligent legend reader.

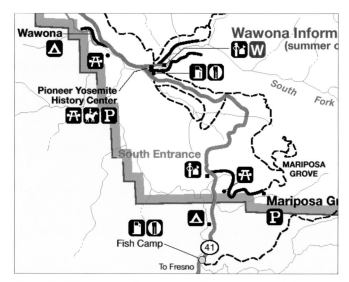

Figure 6.11 *This portion of the Yosemite National Park map uses a standardized set of pictograms. Notice, also, the sparing use of simply designed leader lines from symbols into congested areas of the map.* Source: www.nps.gov/yose/pphtml/maps.html

Figure 6.9 *The visual variable shape is used to represent categories of features at point locations. Local services are represented with three geometric shapes.* Source: ESRI ArcGIS Sample Maps data.

Figure 6.10 *Pictograms use shape to distinguish categories. This small map shows locations for an information office, post office, and library.* Source: ESRI ArcGIS Sample Maps data.

You can use multiple visual variables together. You may find that using hue with shape is a good way to distinguish qualitative differences. The National Park Service map legend *(figure 6.12)* adds green to the pictograms for locations with restrictions. You can also vary size a bit to ensure symbol differences, though you want to avoid an implied order to the symbols. More compact shapes are easier to associate with particular point locations than tall or wide shapes.

You want symbols to be readily identifiable and to look markedly different from each other so readers can identify them without having to look closely at each symbol when they scan the map.

Remember that sometimes the function of symbols is to be seen as groups, not to be read one by one. If all of your symbols are small blue boxes distinguished by different tiny marks within the boxes, none of them are able to stand out from the others. Your readers will not be able to look across the map to see where particular symbols cluster. They will also have a hard time finding the features they seek in the crowd of similar symbols.

It is a good idea to test your symbol shapes and pictograms with map readers unfamiliar with the project to check that your map is easy to understand.

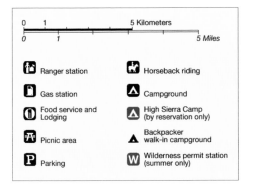

Figure 6.12 *A portion of the map legend for the NPS Yosemite map that identifies the pictograms. Notice the elegantly designed scale bar positioned above the legend.*

Source: www.nps.gov/yose/pphtml/maps.html

Point symbol angle

Symbol angle, or orientation, can be used to vary a point symbol of constant size and shape. A simple use would be to orient a small rectangular symbol vertically, horizontally, and at various angles. The first example in figure 6.13 shows a simple arrow shape at three angles to represent water use. You can see that this visual variable is not particularly effective for this data compared to the same data shown with different symbol sizes in the second example in figure 6.13.

Angle can also be used to vary the internal marks within a symbol, while maintaining a constant shape and size. The first map in figure 6.14 shows different orientations for the two halves within a circle symbol representing qualitative point data. You have also seen this same data represented with different shapes, and the shapes provide more noticeable contrast between features (*figure 6.19, right*).

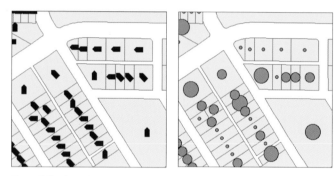

Figure 6.13 *A simple arrow shape at different angles to represent parcel water use (left). The visual variable angle is demonstrated in this map. On the right, the same data is represented more effectively with symbol size.* Source: ESRI ArcGIS Sample Maps data.

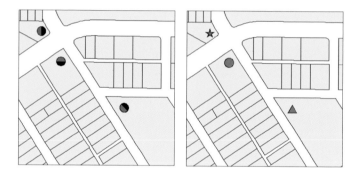

Figure 6.14 *Angles can be set to vary marks within symbols (left), though this may not be as effective as using varied shapes (right).* Source: ESRI ArcGIS Sample Maps data.

The varying orientation of parcels in the first map in figure 6.13 complicates map interpretation. Is symbol angle set relative to the parcel or to the map frame? Angles are set relative to the frame, but a reader may puzzle over the question, reducing the effectiveness of your map.

It is debatable whether the visual variable angle is better suited to qualitative or quantitative data. Three symbol angles could represent different categories of features, though some order seems to be implied by this progression. Angle does not exactly suggest low to high either. Symbols that vary angle can be useful for representing some types of ordered data, such as cyclical and directional phenomena. Weather symbols, for example, make thorough use of angles for wind direction (note the cyan symbols on figure 6.15).

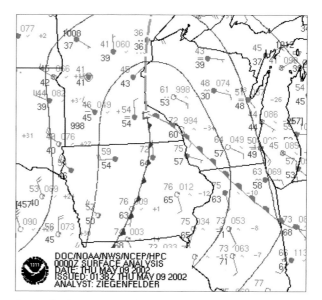

Figure 6.15 *The cyan symbols on this example weather map show wind direction using symbol angle. The "station model" symbols also use many other visual variables, including shape, hue, and arrangement, to report additional weather information.* Source: The National Weather Service as a current surface analysis. Updated maps are available at *www.hpc. ncep.noaa.gov/html/sfc2.shtml*

Line and area symbols

In addition to hue and lightness, the visual variables that may be applied to line and area symbols are the following:

- size
- separation
- shape
- arrangement
- angle

Size is used to construct both proportioned and graduated line symbols. Lines that are dashed and cased combine the visual variables of separation, shape, arrangement, and angle. Line and area symbols use size and separation to reflect ordered differences. Shape, arrangement, and angle of both line and area patterns are used to represent qualitative differences.

Line symbol size

Size is used to represent data values associated with line features by adjusting line widths. Figure 6.16 represents three levels of traffic flow on neigborhood roads.

Figure 6.16 *The visual variable size is used for line features. Traffic flow is well represented with a progression of line widths.* Source: ESRI ArcGIS Sample Maps data.

Just as with point symbol sizes, line sizes may be proportioned or graduated. Proportioned lines are shown at line widths that represent differences in data values. A road that carries ten times more traffic would be shown ten times wider. Graduated line widths rank order lines from low to high, but the line widths are not directly proportioned to data values. Proportioned and graduated lines represent data using the visual variable size.

Proportioned or graduated lines may follow precise routes on a map, such as a road or pipeline. Or, they may be more abstract direction indicators, such as generalized flow lines showing the overall directions of migrations or cumulative measures of travel.

Lines on the two maps below are graduated. Figure 6.17 shows the number of lanes (these lines follow the roads). Figure 6.18 also shows average number of weekly trips from one town to other nearby locations (these lines run between points without following roads).

Figure 6.17 *Line size is graduated to represent the number of road lanes. The graduated lines follow the roads in this example.* Source: Prince George's County, Maryland.

Figure 6.18 *Line size is graduated to represent the average number of trips from one town to nearby locations in this hypothetical example. The lines summarize multiple routes and represent generalized directions.* Source: Prince George's County, Maryland.

Line symbol patterns

There are multiple line patterns available for map design. The primary options for line symbols in map design are dashing and casing.

Dashes add pattern to a line, and differences in separation and length of the dash pattern are used to create different symbols. Wider spacing between dashes and longer dashes create coarse textures.

As dash patterns become more complex, they also make use of shape and arrangement. An international boundary line that combines short and long dashes is a combination of separation, arrangement, and shape characteristics. Figure 6.20 shows the same shape with different arrangements. Lines can also be built from differently shaped marks, such as a string of dots that contrasts with a string of crosses, as shown in figure 6.21.

Figure 6.19 *An example use of separation as a visual variable for dashing lines.* Source: ESRI ArcGIS Sample Maps data.

Figure 6.20 *Arrangement of marks along a line symbolizes different road types.* Source: ESRI ArcGIS Sample Maps data.

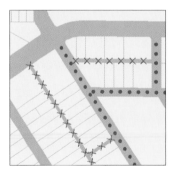

Figure 6.21 *A wide variety of shapes may be used to define line symbols. Circles and crosses distinguish two road types in this example map.* Source: ESRI ArcGIS Sample Maps data.

The angle of the marks within a line symbol can be used to indicate line category. Hatching is created when a pattern is at an angle to the line. In figure 6.22, some hatches are set perpendicular to the line direction and some are at an angle. The cased alleys are included in the example because you can interpret line casing as an extreme angle setting (though you do not create line casings the same way you create hatching). A casing pattern runs parallel with the line direction.

Figure 6.22 *The visual variable angle is used to vary line patterns in this example map.* Source: ESRI ArcGIS Sample Maps data.

Casing is a commonly used and versatile cartographic symbol option for lines. Casing functions just like halos do for text—it helps increase line visibility over multiple backgrounds. It can also be used to create a wide symbol that contrasts with its background but is not overly bold (that is, it does not overwhelm the visual hierarchy among the map symbols).

Casings are lines on both sides of the symbol, and they may be thin or thick. If the two lines that build a line symbol are similar in width, then the casing is thin. The example portion of a road map (*figure 6.23*) shows a variety of cased lines.

The legend is labeled with the casing characteristics (rather than numbers of road lanes, the mapped data). The 3-point pink line nested in the 4-point black line produces a 0.5-point casing on both sides of the line (splitting the 1-point difference between 3- and 4-point lines). The same 3-point line nested in an 8-point black line produces a thicker 2.5-point casing.

More than two layers can be used to build cased lines. The widest line on the map has three lines layered to create lanes, as shown in segments of the ArcMap symbol property editor below.

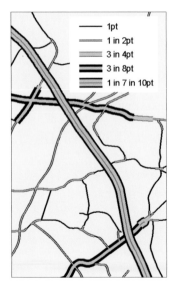

Figure 6.23 *Example cased lines that have varied overall widths, varied casing widths, and multiple lanes.* Source: ESRI ArcGIS Sample Maps data.

Figure 6.24 *The layers that build three example cased lines from the above map: 3 in 4 point, 3 in 8 point, and 1 in 7 in 10 point. Note the symbol preview above each set of layers.*

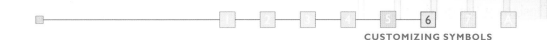

In figure 6.23, notice that some casings interrupt intersecting lines and others do not. You can control the interaction between cased lines in GIS. In ArcMap, this design detail is set using join and merge toggles *(figure 6.25)*. In design software, layering and order settings within layers help to control the interaction of lines and casings.

Wide lines can carry a pattern that involves the whole suite of visual variables that may be applied to lines: separation, shape, arrangement, and angle. Add color variables and you can produce an almost endless variety of line patterns. But that does not mean that you should use intricate line symbols in their fullest complexity; remember, don't stump your reader.

Figure 6.25 *Join and merge toggles in the advanced-symbol-level drawing options in ArcMap control the way different cased line features intersect or break other lines.*

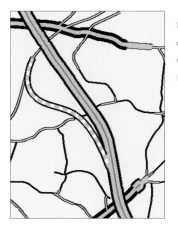

Figure 6.26 *A line that is dashed and cased is an example of using multiple visual variables for line pattern.* Source: Prince George's County, Maryland.

A cased line will indicate a different kind of feature than uncased or dashed lines. For example, a cased highway will be distinguished from a dashed county line. You can even case a dashed line (as shown in figure 6.26 for a proposed highway off-ramp). Combinations of casing, dashing, and width, along with other visual variables, will establish a hierarchy of lines.

Area patterns

A wide array of area patterns may be used on maps. They can be quite literal in their design, such as the repeated pattern of small tufts of grass and reeds used to represent a swamp *(figure 6.27)*. Or they can be completely abstract, such as a crosshatch of evenly spaced lines across an area.

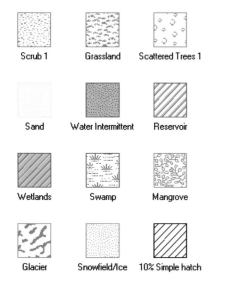

Figure 6.27   *A selection of pattern styles made available in ArcMap.*

Choose area patterns by paying attention to the visual variables used to build them. You want to choose and customize area patterns so they obviously represent the logical relationships within the data, just as you did with color symbols.

Use textures that are coarse and fine to represent hierarchy *(figure 6.28, left)*. Loosely spaced textures represent low values and closely spaced textures represent high values. Similarly, a fill of larger shapes will indicate a higher data value than a fill of small shapes with the same spacing *(figure 6.28, right)*.

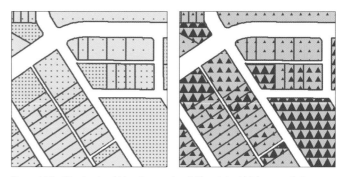

Figure 6.28   *The visual variables of separation (left) and size (right) are applied to area symbols to establish hierarchy.* Source: ESRI ArcGIS Sample Maps data.

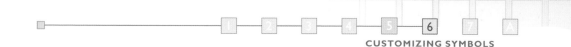
Use the shapes of elements within a pattern to represent qualitative differences within the data *(figure 6.29, left)*. A fill of small circles will indicate a different kind of feature than fills of crosses or stars. More literally, a fill of small tree drawings may contrast with a pavement pattern to distinguish park from parking. Angle *(figure 6.29, right)* and arrangement *(figure 6.30)* can also be used with area patterns to indicate qualitative differences.

As with points and lines, visual variables can be used together to increase the contrast between area fills. Alter lightness to enhance hierarchy and use hue to enhance qualitative differences. Figure 6.31 combines hue, lightness, arrangement, angle, separation, shape, and saturation to create a range of area patterns.

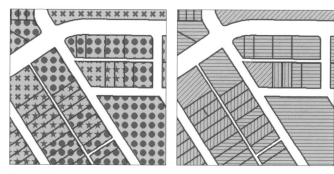

Figure 6.29 *The visual variables of shape (left) and angle (right) are applied to area symbols to establish categories.* Source: ESRI ArcGIS Sample Maps data.

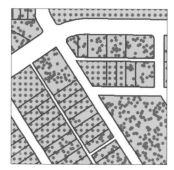

Figure 6.30 *ArcMap offers two arrangement settings (grid and random) that can be useful for representing qualitative differences between areas.* Source: ESRI ArcGIS Sample Maps data.

Figure 6.31 *This map combines hue, lightness, arrangement, angle, separation, shape, and saturation to create high-contrast area patterns for different land uses.* Source: ESRI ArcGIS Sample Maps data.

Wrapping up visual variables

Can you envision patterned point symbols that vary in separation? Do you recall the look of line symbols that vary in arrangement? With eight visual variables and three types of features, you have twenty-four basic ways to vary symbols for representing map data.

Because there are many combinations, this chapter concludes with a pair of tables that provide you with a summary view of point, line, and area features represented with symbols that vary in color, size, shape, and pattern. The visual variables organized in these tables are the following:

- hue
- lightness
- saturation
- size
- shape
- separation
- arrangement
- angle

Visual variables for ordered data

Lightness, saturation, size, and separation are the visual variables well suited to representing ordered data (either rank-ordered data or numerical amounts) *(table 6.1)*. These visual variables establish hierarchies among features. Additional visual variables used for symbolizing quantitative data are perspective height, transparency, and crispness (focus).

Visual variables for qualitative data

Hue, shape, arrangement, and angle are the visual variables well suited to categorizing features *(table 6.2)*. They represent qualitative differences that are not ordered. Symbol angle may also be useful for representing some types of quantitative data, such as direction.

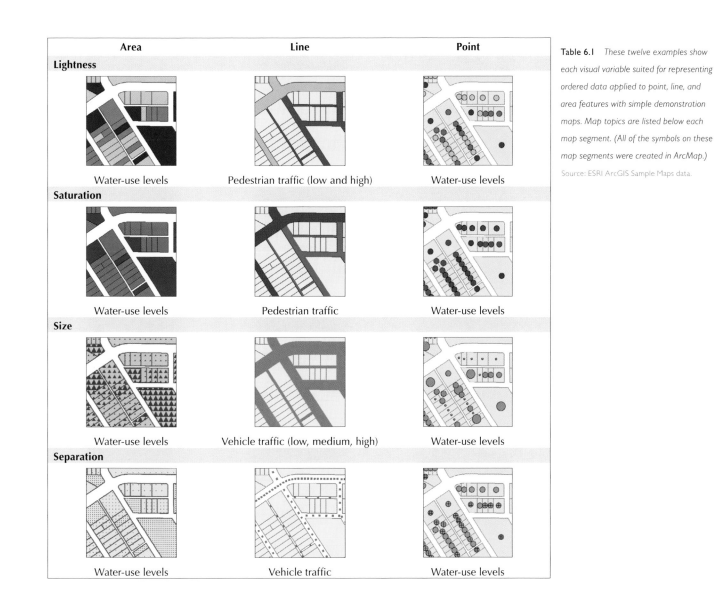

Area	Line	Point
Lightness		
Water-use levels	Pedestrian traffic (low and high)	Water-use levels
Saturation		
Water-use levels	Pedestrian traffic	Water-use levels
Size		
Water-use levels	Vehicle traffic (low, medium, high)	Water-use levels
Separation		
Water-use levels	Vehicle traffic	Water-use levels

Table 6.1 *These twelve examples show each visual variable suited for representing ordered data applied to point, line, and area features with simple demonstration maps. Map topics are listed below each map segment. (All of the symbols on these map segments were created in ArcMap.)*
Source: ESRI ArcGIS Sample Maps data.

Table 6.2 *These twelve examples show each visual variable suited for representing qualitative data applied to point, line, and area features with simple demonstration maps. Map topics are listed below each map segment. (All of the symbols on these map segments were created in ArcMap.)*

Source: ESRI ArcGIS Sample Maps data.

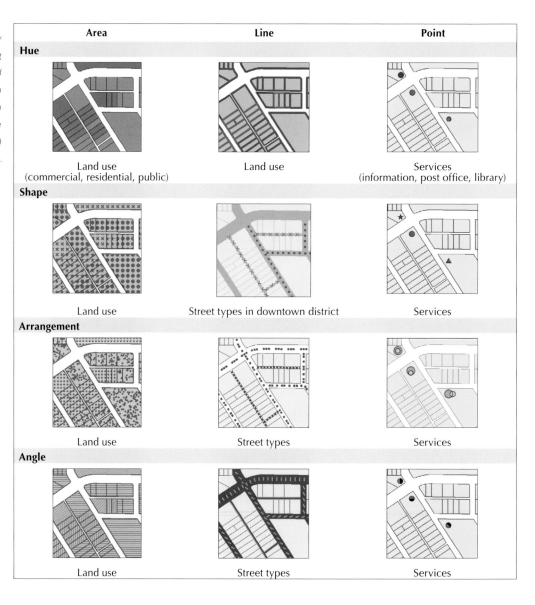

	Area	Line	Point
Hue	Land use (commercial, residential, public)	Land use	Services (information, post office, library)
Shape	Land use	Street types in downtown district	Services
Arrangement	Land use	Street types	Services
Angle	Land use	Street types	Services

7

BEYOND DEFAULT MARGINAL ELEMENTS

CHAPTER 7

BEYOND DEFAULT MARGINAL ELEMENTS

Concise and well-placed words communicate information clearly. Mapmakers design titles, legends, notes and other marginal elements to help map readers understand and remember their maps. The care with which you compose and finish the elements that surround and support a map make it work.

Building better maps through wording and placement requires revisiting map layout techniques from chapter 1. It means providing information without disrupting the visual hierarchy that emphasizes the main map content. It also means paying attention to nuances and details of order, placement, and interrelationships in content. Excellent map design is definitely detailed work, but the more you know about the design tools you have in GIS and the cartographic conventions for marginal elements, the quicker you will be able to complete professional, readable map designs.

Designing better maps means finishing with the following:

▫ wise wording, attending to hierarchy in text content, description of mapped calculations, logical line breaks and spacing

▫ well-designed marginal elements, including map legends and indicators of scale and direction

Wise wording

When you begin creating a map using GIS, the wording of basic text elements is set at defaults. The legend may be labeled "Legend" and raw layer names will display as headings. Symbols will be labeled with attributes from data tables. These text elements are fine as cues to what is mapped when you are first exploring the spatial distributions of your data. But when you get serious about map design you will want to edit text elements.

Your goal is to communicate the map content clearly using a hierarchy of detail. You also want to refine labels so spacing within and between lines of text conveys clear associations with other map elements. This section asks you to bring a critical eye to the content and arrangement of your map title, subtitles, legend titles, and notes.

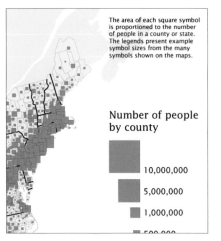

Figure 7.1 *Map segment showing a hierarchy of text wording. The map title (not shown) is "Population Distribution, 2000."* Source: Brewer and Suchan. 2001. *Mapping Census 2000: The Geography of U.S. Diversity.* U.S. Census Bureau.

Hierarchy in text content

In addition to the labels within the map body, you can add a variety of marginal text elements to a map. The map usually has a title and legend title. You may also add subtitles and notes to explain the data that is mapped or the purpose of the map to your reader.

Marginal text on maps should be succinct with minimal punctuation. An example of hierarchy in map text is listed below:

1. Short title:

Population Distribution, 2000

2. Legend title elaborates:

Number of people by county

3. Note completes detail:

The area of each square symbol is proportioned to the number of people in a county or state. The legends present example symbol sizes from the many symbols shown on the maps.

The title (1) is a straightforward statement of the overall map topic. More detailed information about the map content is stated in the legend title (2), and further explanation is offered in the note (3). The legend and note are also shown in figure 7.1. In this example, the note offers assistance in interpreting symbols for two maps on the page. Sentences and full punctuation are used only in the smallest text element, the note.

Figure 7.1 has fairly straightforward content. But how do you handle wording for more complex concepts that can be mapped? It might seem straightforward to require that a map title list who, what, where, and when, and that both the numerator and denominator of a mapped ratio be made clear. In reality, these details can produce an impenetrable title.

Here is an accurate but poor title for the map shown in figure 7.2:

A map showing the distribution of the percent of people indicating one or more races including American Indian and Alaska Native who are under age 18 in 2000 by county in the United States prepared using Census 2000 Redistricting Data

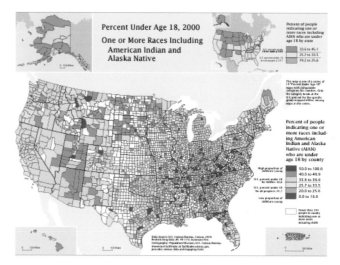

Figure 7.2 *Example map layout. Segments of this map are shown enlarged in the figures that follow.* Source: Brewer and Suchan. 2001. *Mapping Census 2000: The Geography of U.S. Diversity.* U.S. Census Bureau.

The potentially unwieldy list of who, what (numerator and denominator), where, when can be parceled out among the title, subtitles, legend, and notes to help map readers understand the map.

For individual maps, where and when can often be listed in a subtitle, one step down from the title in the hierarchy of text on the map. When you are preparing a series of maps, where and when are often covered in the introductory material for the group of maps. They need not be repeated in the title of each map. All of the U.S. map examples shown in this chapter (figures 7.1–4, 6, 9, 13, and 27) are from an atlas of Census 2000 population data published by the Census Bureau. In the examples, "United States" is not included in map titles *(figure 7.2)* because the whole series is clearly about the United States.

What and who for a map can range from obvious to convoluted. The details of numerators and denominators of ratios may be difficult to understand. They can be explained more fully in the legend title that augments a summary map title. Who can also be complex if, for example, a subset of a group or multiple groups in a population are being described.

Details like source information should not appear in a title and should be reported in a note in small text that is not prominent on the map. Other content suited to notes in small text include author information, explanations of map calculations, and tips for map reading.

The shortened title used for the example atlas map in figure 7.2 is

Percent Under Age 18, 2000

One or More Races Including American Indian and Alaska Native

The legend title in figure 7.3 (enlarged from figure 7.2) picks up many of the details from the inappropriately long title proposed at the beginning of this example. The annotations on the left side of the legend further assist the reader in understanding the map.

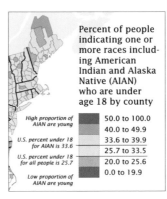

Percent of people indicating one or more races including American Indian and Alaska Native (AIAN) who are under age 18 by county

High proportion of AIAN are young

U.S. percent under 18 for AIAN is 33.6

U.S. percent under 18 for all people is 25.7

Low proportion of AIAN are young

50.0 to 100.0
40.0 to 49.9
33.6 to 39.9
25.7 to 33.5
20.0 to 25.6
0.0 to 19.9

Figure 7.3 *The legend title and legend annotations from the example map in an atlas of Census 2000 population data published by the U.S. Census Bureau*
Source: Brewer and Suchan. 2001. U.S. Census Bureau.

Notes from various places on the map are pulled together in figure 7.4 to show the level of content that belongs in the smallest text on the map. These details should not be in the map title or legend title. The title needs to be straightforward enough to invite the reader to be interested in your map. Offer them details in the fine print, rather than distracting them from your map content with too many words.

The previous titles for the Census Bureau atlas map are somewhat wordy but meet needs for clarity. Race groups are presented in two ways in the atlas (including and excluding people who indicated more than one race on their census form). Simpler titling would suit the same content presented in a different context. An alternative set of title, legend, and note wordings below is less specific in race group naming and allocates more of the detail to notes *(figure 7.5):*

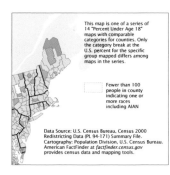

Figure 7.4 *Additional notes from the "Percent Under Age 18" map (moved together to show them in one figure) from an atlas of Census 2000 population data published by the U.S. Census Bureau.* Source: Brewer and Suchan. 2001. U.S. Census Bureau.

1. Alternative title:

Native American Children, 2000

2. Legend title:

Percent of Native American population who are under age 18 by county

3. Explanatory note:

The Native American population mapped includes both American Indian and Alaska Native (AIAN) groups. This population includes people indicating their race as AIAN alone and those indicating AIAN in combination with other races.

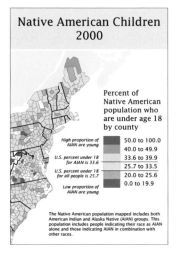

Figure 7.5 *An alternative wording for the map of Census 2000 data shown in figures 7.2, 7.3, and 7.4. The map title, legend title, legend annotations, and note have been collected into a corner of the page. This wording may not conform with language approved within the agency for describing race groups.* Source: Adapted from Brewer and Suchan. 2001. U.S. Census Bureau.

Describing mapped calculations

Describing a calculation in few words can be harder than you expect. You might end up suggesting a different map topic than you intend if you do not pay close attention to the wording.

The map below is related to the American Indian and Alaska Native map discussed in the previous section. This example seems to be simpler for wording the legend title because Census 2000 ethnicity data is categorized into only two groups: Hispanic and Not Hispanic (ethnicity is a different categorization than race). The wording used in the atlas map is shown in figure 7.6.

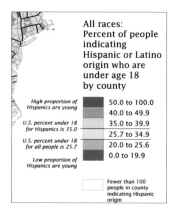

Figure 7.6 *Legend wording used for the map showing the distribution of the young Hispanic population.* Source: Brewer and Suchan. 2001. U.S. Census Bureau.

The legend caption chosen for the atlas map is fairly wordy. Why, you might wonder. Couldn't the words be cut in half and still get the idea across? Some wording options are listed below with notes to alert you to potential misinterpretations.

1. Correct but wordy:

Percent of people indicating Hispanic or Latino origin who are under age 18 by county

2. Ambiguous:

Percent Hispanic under 18 by county

Instead of the intended meaning (graph on left in figure 7.8), this option could suggest percent of total population (right in 7.8) or percent of total under-18 population (middle in 7.8).

3. Confusing:

Under 18 Hispanic percent by county

This option might mean percent of the total under-18 population who are Hispanic (middle graph in figure 7.8).

4. Shorter but OK:

Percent Hispanic who are under 18 by county

This caption should be interpreted accurately (left graph in figure 7.8).

Simple graphs *(figures 7.7 and 7.8)* demonstrate the nuances of potential meaning expressed by the four legend captions above. You do not want to confuse your map reader with the text you provide, and you also do not want them to misinterpret your text because it is excessively brief.

Be sure your map text says what you mean to say. As you work on a map, ask others to describe the map to you to check that you have not gone off-track in your wording. Do not describe the map to them; have them describe it to you.

If you are having difficulty balancing brevity and accuracy for title and legend wording, consider adding a note that explains the calculation. You may include the formula for the calculation in a note. A portion of the note for a diversity map is shown below as an example of this strategy. Readers interested in the math behind a mapped distribution will gain more from the map if they are sure of what it presents. Readers who are less interested in the math can ignore the details because they are listed in a small note.

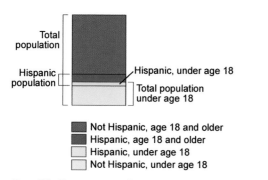

Figure 7.7 *The map shown in figure 7.6 presents the ratio of the yellow segment to the sum of the green and yellow segments of the population in this bar graph.*

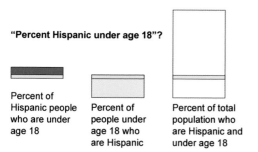

"Percent Hispanic under age 18"?

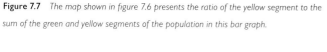

Percent of Hispanic people who are under age 18

Percent of people under age 18 who are Hispanic

Percent of total population who are Hispanic and under age 18

Figure 7.8 *The ambiguous caption presented above, "Percent Hispanic under age 18," could steer a map reader to different interpretations, shown here with explanatory graphics (figure 7.7 defines the colors used).*

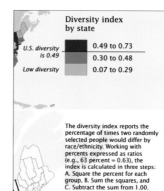

Figure 7.9 *The beginning of a note that describes the diversity calculation used for a map in an atlas of Census 2000 population data published by the U.S. Census Bureau.* Source: Brewer and Suchan. 2001. U.S. Census Bureau.

Attending to line logic

After you have decided on your map wording, you need to attend to the nuances of where lines break and spacing between text elements. Breaks and spaces take over some of the roles of sentence structure and punctuation that are not fully used for map text. You learned about the importance of proximity for map interpretation in chapter 1. This issue is addressed in greater detail here.

A legend from an example you have seen before is shown in figure 7.10, but it has been revised with thoughtless breaks and spacing. Do you know what this title means? Something about "transportation and land," "using Prince George's County," and "Maryland land use" perhaps? Which symbols do "Institution" and "Parkland" label?

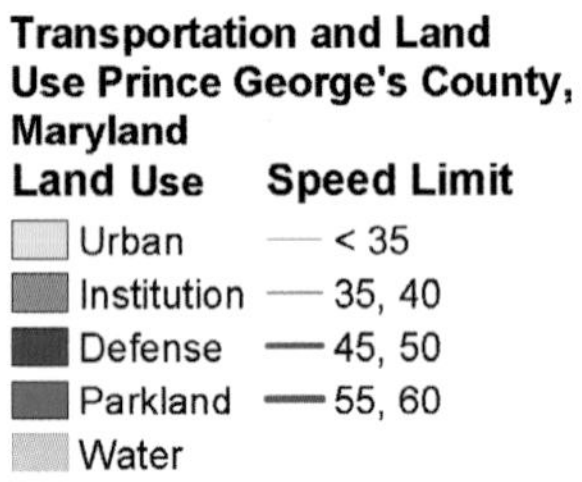

**Transportation and Land
Use Prince George's County,
Maryland**
Land Use   **Speed Limit**

▢ Urban   —— < 35
▨ Institution   —— 35, 40
▮ Defense   —— 45, 50
▨ Parkland   ▬ 55, 60
▨ Water

Figure 7.10   *Text for a map of Prince George's county, Maryland, with poor line breaks and compressed spacing between elements.* Source: Prince George County, Maryland, PG Country.

The revised example in figure 7.11 shows the same content with better breaks, a hierarchy of type sizes, and more spacing between elements. In the title, the two words in "Land Use" are now on the same line. "Maryland" has been pulled into the subtitle. The legend titles are separated from the subtitle so you can tell they are clearly part of the legend. Additionally, a larger gap between the two columns of the legend clarifies that the land-use labels apply to the symbols on the left.

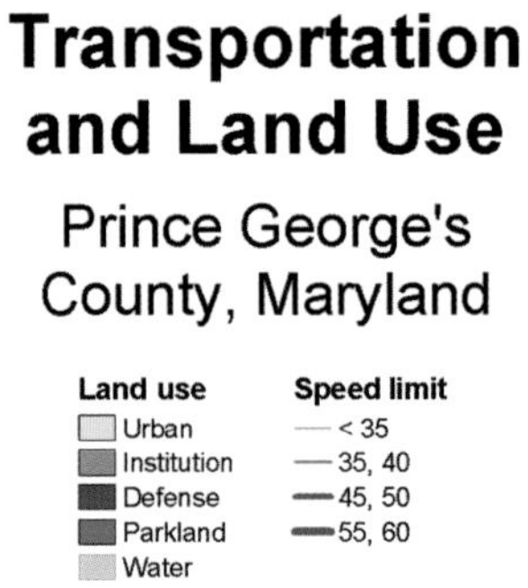

Transportation and Land Use

Prince George's County, Maryland

Land use   **Speed limit**
▢ Urban   —— < 35
▨ Institution   —— 35, 40
▮ Defense   ▬ 45, 50
▨ Parkland   ▬ 55, 60
▨ Water

Figure 7.11   *Spacing and line breaks are used to clarify map content for the Prince George's County map.* Source: Prince George County, Maryland, PG Country.

Note that the legend in figure 7.11 is not titled "Legend." It is better practice to use legend titles and headings that describe legend content. After all, it is quite obviously a "Legend," so that label is superfluous.

The legend area of a Census 2000 atlas map is shown in figure 7.12 without spacing between text elements or a hierarchy of sizes and styles. The annotations on the left are particularly difficult to interpret.

The annotations to the left of the legend are each separated by spaces and are closer to the portion of the legend to which they refer than to anything else. The white legend box is moved down so the "Hispanics are young" phrase does not seem to refer to it. The legend is an improvement on the one in figure 7.12.

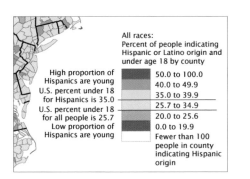

Figure 7.12 *Legend text without careful spacing or hierarchy for the Census 2000 atlas map showing the distribution of the young Hispanic population.* Source: Brewer and Suchan. 2001. U.S. Census Bureau.

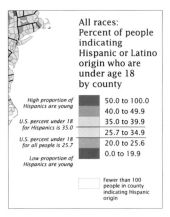

Figure 7.13 *Legend text spacing and hierarchies used in an atlas of U.S. Census 2000 population data published by the U.S. Census Bureau.* Source: Brewer and Suchan. 2001. U.S. Census Bureau.

The corresponding legend in the atlas *(figure 7.13)* uses spacing and type hierarchies to separate text elements and clarify which parts of the legend they refer to. The legend title is larger than the other text, asking the reader to attend to its content before the details below it.

You should break lines to retain the logic of phrases within the text. You do not need to use the entire length of the available space before breaking to the next line.

You should use spaces carefully as well. Keep lines in a text block closer to each other than to other text elements on the map. Likewise, put a text element closer to the element it labels than to anything else on the map. This is much the same logic you learned for placing labels within the body of the map. You want to consider positioning for marginal text elements as carefully as you do for map labels.

Well-designed marginal elements

Legends, scales, and north arrows are basic marginal elements on maps. Legends present information that lets the map reader understand symbols and interpret the map. In this section, you will learn the basic legend components for a variety of standard thematic mapping methods. Scale bars and direction indicators can be customized to control visual hierarchy and give your map a distinctive look. Be aggressive about editing the appearance of these other marginal elements as well.

Map legends

Map legends explain map symbols. The basic types of thematic maps have fairly conventional legend content. GIS software provides only some of this content within legend tools. This section provides an overview of what you should be aiming for as you adjust default legends for the following types of thematic maps:

- choropleth
- qualitative area fills
- dot (density)
- contours and other isolines
- proportioned symbols
- segmented symbols

Choropleth—A legend for a choropleth map states the data ranges for each color or pattern used. Choropleth maps are produced using graduated colors as area fills that may be organized as sequential or diverging schemes.

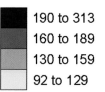

Incidents per 100,000 people

190 to 313
160 to 189
130 to 159
92 to 129

Figure 7.14 *A choropleth map legend.*

As you label a choropleth map legend, you need to decide whether to do the following:

- round numbers for breaks or within labels

- increment labels (e.g., 0–10, 10–20 … or 0–9, 10–19; this issue is linked to rounding)

- use the word "to" or a dash within ranges (this may depend on whether the data includes negative numbers)

- label breaks between classes with single numbers rather than labeling class ranges

- order classes with the highest numbers at the top (like the vertical axis on a graph) or at the bottom of the legend

- label ranges with the actual values represented by the symbol, creating gaps between ranges

- use the true maximum and minimum in the dataset to label ranges or use statements such as "fewer than 100 people" or "more than 150 percent" for extreme ranges

- add annotations that elaborate on the meaning of ranges or breaks to assist map reading

Qualitative area fills—Maps that use qualitative area fills provide a descriptive label for each legend color or pattern. Creating logical groupings of related categories assists map reading.

Figure 7.15 *A qualitative area map legend.*

For area symbols, the legend boxes should present colors or patterns in the same way they appear on the map. If map polygons have outlines, use the same outline color and weight in the legend. If there are no lines between colors on the map, consider presenting the symbols that way in the legend *(figure 7.15)*. If colors are seen immediately adjacent to each other on the map, show them that way in the legend as well *(figure 7.14)*.

Dot (density)—At their simplest, dot map legends define the amount that one dot represents. Dot map symbols are intended to show density (the dots are not counted and they do not show precise location). A dot map is improved if you include example densities in a set of legend boxes in addition to defining the dot value. You will need to manually construct the density boxes for the legend.

One dot represents 300 people

Figure 7.16 *The minimum legend for a dot map defines the meaning of a single dot.*

Sample densities in people per square km

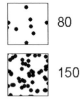
80

150

300

Figure 7.17 *A complete legend for a dot map includes example densities.*

Each square represents 100 square km

Contours or other isolines—The simplest legend on a contour map or other isoline map is a statement of the isoline interval. If the isolines are augmented through use of index lines, hachures for closed depressions, or supplemental lines, you may choose to identify the meaning of these different line symbols in the legend. If color fills are used between isolines to enhance the map pattern, present those colors in a legend. You may label the colors with data ranges or label the isolines that mark breaks between colors. Example isoline legends are shown in figures 7.18, and 7.19.

Isoline interval is 200 meters

Figure 7.18 *A minimum legend for an isoline map defines the isoline interval.*

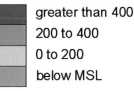

Elevation in meters above mean sea level

greater than 400
200 to 400
0 to 200
below MSL

Elevation in meters above mean sea level

400
200
0

Figure 7.19 *A map with color fills between contours (hypsometric tints) includes a legend that identifies the range for each color (left). Alternatively, the legend may define the breaks between colors rather then the data range for each color (right).*

Proportioned symbols—Proportioned point symbol legends present a set of example symbol sizes to demonstrate the amounts that map symbols represent. The best strategy is to show the smallest and largest symbol sizes seen on the map, so the map reader can interpolate between these extremes. Include one or more intermediate size symbols to assist map reading *(figure 7.20)*.

Number of people

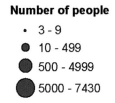

- 3 - 9
- 10 - 499
- 500 - 4999
- 5000 - 7430

Figure 7.21 *Legends for graduated symbol maps are much like choropleth map legends (graduated color maps). They define the data range for each symbol size.*

Number of people

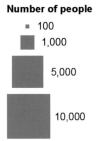

- 100
- 1,000
- 5,000
- 10,000

Figure 7.20 *Legends on proportioned point symbol maps show the largest and smallest symbols and a range of sizes between these extremes to aid interpolation.*

You may use symbols rounded to even amounts that approximate the data range rather than exact minimum and maximum symbol sizes in the point symbol legend. If the largest symbol on the map is too large to be repeated in the legend, consider labeling the amount next to that symbol (and a few other very large symbols) on the map and use a more manageably sized large symbol in the legend.

If you use graduated rather that proportioned point symbols, be sure to clarify that sizes do not represent precise data values by listing data ranges for each symbol in the legend *(figure 7.21)*. Follow choropleth legend conventions for graduated symbol maps.

Segmented symbols—Segmented symbols, such as pie charts and bar graphs, require custom legend design. Construct a generic looking symbol and label each segment. If the segmented symbols also vary in size, create a two-part legend that uses a different set of example symbols to define sizes.

Bar height: Number of employees by township

1000
500
200
100

Figure 7.22 *Legends for multivariate symbols, such as segmented bar symbols, should step the reader through the meaning of each symbol component. This legend explains bar heights and bar segment colors in two steps.*

Bar segments: Percentage employed by sector

Government
Private

100%
75
50
25
0

There are many other map symbols that you may find useful. You should customize legend titles and headings, labels, and layout to best describe the meaning of these symbols. Do not feel, though, that every single symbol must be in a legend. If your base information symbols (roads, water, and boundary lines, for example) are fairly obvious in their meaning, you may leave those symbols out of the map legend.

When you need to customize a legend beyond the basic settings that can be made for legend properties, you need to sever the link between the data and the legend. The legend then becomes a graphic that does not change in response to symbol changes on the map. There are four basic strategies for customizing legends:

1. Insert a legend, convert it to graphics, ungroup the graphics, and edit individual elements.

2. Add drawn elements and additional text to an inserted legend.

3. Convert symbols to graphics and copy and paste examples from the map to build a legend.

4. Construct the entire legend manually with text and drawing tools.

With any of these strategies, you will need to manually update legend elements if you change the map. For example, if you change symbol colors and sizes, you will need to manually make corresponding changes in the legend graphics. If you change the data ranges represented by map colors or other symbols, you will need to manually edit the text information in the legend. To minimize the manual work required to ensure a match between map and legend symbols, legend customization should be done at the very end of the map design process.

The example in figure 7.23 shows how many elements a legend breaks into when you convert it to graphics and ungroup the elements. Each block of color, each line symbol, and each label becomes an editable element. You can move these elements, change color and line characteristics, and edit the text to better suit the map.

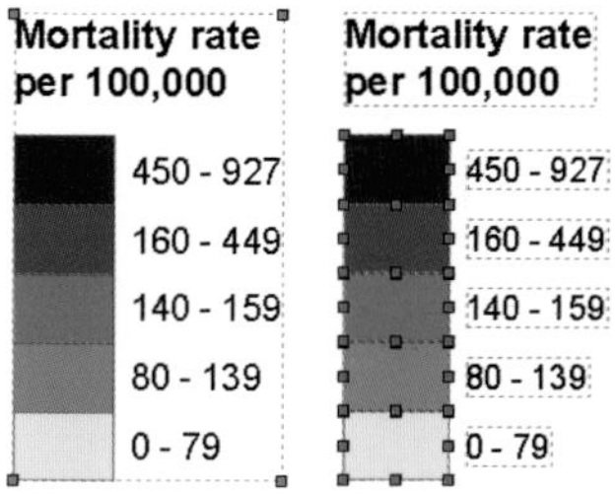

Figure 7.23 *Example choropleth map legend (left) converted to graphics (right).*

Annotations added to a map legend *(figure 7.24)* clarify the meaning of groups of map classes and of particular breaks between colors. The underlying legend remains live; it will update if classes or colors are changed. The annotations, however, will need to be manually edited because they are not linked to the map data.

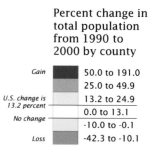

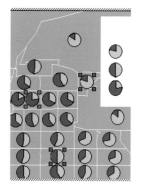

Figure 7.24 *Added text, to the left, explains legend classes. Two meaningful breaks are noted and two general groupings of classes are described as Gain and Loss for this atlas map. The black lines across two breaks are also added drawings.* Source: Brewer and Suchan. 2001. U.S. Census Bureau.

Figure 7.25 *The three selected symbols were copied and pasted into the legend area being built in the upper right of the frame. The map shows the relative proportion of family (pink) and nonfamily (red) households by census tract in part of Salt Lake City.* Source: Census 2000.

Another handy way to construct a custom legend is to convert map features to graphics and then copy and paste example symbols to compile a legend. This is especially useful for multivariate chart symbol that are difficult to redraw.

The example in figure 7.25 shows pie charts converted to graphics. Three example symbols are copied into a legend area to begin preparing a detailed explanation of symbols. Remember, converting either the legend or map symbols to graphics breaks the link to the data, so do this only when you are at the very end of the map design process.

The two-variable legend in figure 7.26 was manually constructed with ArcMap drawing tools. Lines and boxes were drawn, colors set, and text placed to build this customized legend.

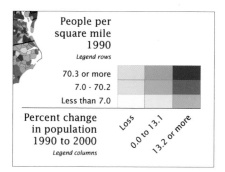

Figure 7.26 *A manually constructed legend for a two-variable map from the Census 2000 atlas.* Source: Brewer and Suchan. 2001. U.S. Census Bureau.

The ArcMap legend tools get you part of the way to a custom legend. You will sometimes want to copy, augment, or draw parts of your legend to explain customized map symbols.

Scale and direction indicators

A scale and north arrow are basic marginal elements used on most, but not all, maps.

Map scale is communicated three ways: as a graphic bar scale, a verbal scale, and a representative fraction (absolute scale). An example of each scale type is shown in figure 7.27. The graphic bar scale remains most accurate in the dynamic environments in which we look at maps because it resizes as map size varies with zooming, screen resolution, and reproduction.

1 centimeter equals 0.685 kilometers

1:68,521

Figure 7.27 *Three forms of map scale. From the top, graphic scale, verbal scale, and representative fraction.*

As discussed in chapter 1, detailed scale bars are useful for maps that will be used for distance measurement. Only a general indication of distances is usually needed on thematic maps that present statistical data, so a simple scale bar is more suitable.

A good scale bar presents rounded units that the map reader can easily use. The example below shows a scale bar with poorly chosen units and one with more useful rounded units of 5 and 10 miles.

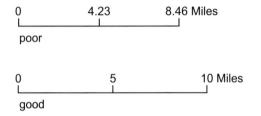

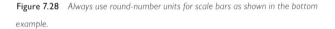

Figure 7.28 *Always use round-number units for scale bars as shown in the bottom example.*

Be aware that on small-scale maps, such as a map of the world, scale varies across the map. A single scale bar will not be an accurate representation of scale for much of the map.

Scale bars can also be converted to graphics and edited, though they will no longer update if the map scale is changed. Make these types of edits only after the map extent and scale has been finalized.

GIS software offers a wide variety of north arrow designs. The north arrow you choose or design can be a distinctive element of a map, acting much like a logo for your work. But a north arrow, is rarely the most important feature on a map, so keep that in mind when designing this element. Do not let it get so large or elaborate that it draws attention away from the map content.

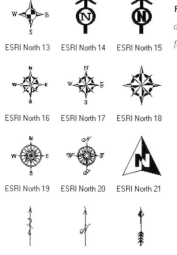

Figure 7.29 *Example north arrow designs available in ArcMap that range from simple to ornate.* Source: ESRI ArcMap.

ESRI North 13   ESRI North 14   ESRI North 15

ESRI North 16   ESRI North 17   ESRI North 18

ESRI North 19   ESRI North 20   ESRI North 21

Once you select a north arrow style, there are few ways to customize it. You can set the north arrow size and adjust its background. If you want to do further customization, you will need to draw an arrow from scratch.

On small-scale maps, north may be in many directions. When lines of longitude converge within a map projection, a north arrow will always be wrong for some parts of the map *(figure 7.30)*.

Do not use a north arrow when the direction of north varies across the map. For example, maps of the United States or Canada created using customary conic projections should not include a north arrow because longitude lines converge toward the pole. It is also fine to omit the north arrow from maps of a place familiar to the intended audience.

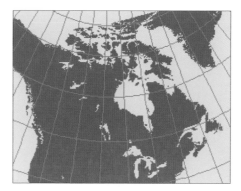

Figure 7.30 *Do not put a north arrow on a map like this, because north is many different directions within the map.* Source: ESRI Data and Maps CD/DCW data.

When north is multiple directions, replace the north arrow with the graticule to indicate direction. The graticule is a grid of lines of longitude and latitude and thus makes a fine directional indicator. As with north arrow design, the graticule is supporting information, so it should have a subdued design that does not compete with map content.

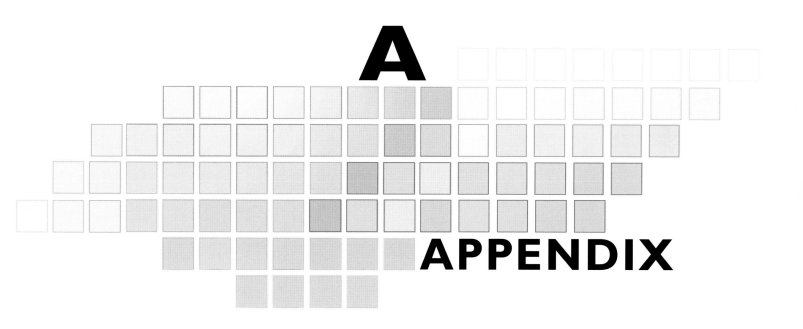

A

APPENDIX

ColorBrewer

ColorBrewer is an online tool that offers color schemes for thematic mapping. Colors are specified in RGB, CMYK, hexidecimal (for Web color), CIE Lab (a perceptually scaled system), and HSV for ArcView® 3.x (which uses nonstandard HSV dimensions). Mark Harrower designed the interface and programmed the Flash Web site at *www.ColorBrewer.org* that delivers the color schemes. The schemes were designed by Cindy Brewer. Color-Brewer is based upon work supported by the National Science Foundation under Grant No. 9983451, 9983459, and 9983461, under the Digital Government program.

The CMYK and corresponding RGB values (converted in Adobe Illustrator) are listed in this appendix. In addition to ordered sets of colors (as they would appear in a map legend), each scheme is represented with small, intermixed polygons to the right of the legends. These map-like strips assist you in evaluating whether colors in a scheme look different enough to be identified when they are small and positioned within a more complex background.

The diagrams in this appendix were first published in

Brewer, Cynthia A., Geoffrey W. Hatchard, and Mark A. Harrower. 2003. ColorBrewer in Print: A Catalog of Color Schemes for Maps. *Cartography and Geographic Information Science.* 30(1): 5–32.

They are reprinted with the permission of the American Congress on Surveying and Mapping.

An example

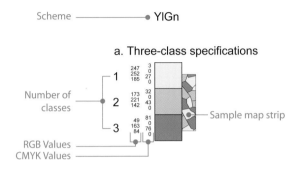

Scheme ──────→ YlGn

a. Three-class specifications

Number of classes

Sample map strip

RGB Values
CMYK Values

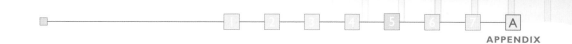
RGB and CMYK specifications for nine sequential schemes with hue transitions in ColorBrewer

Columns: **YlGn**, **YlGnBu**, **GnBu**, **BuGn**, **PuBuGn**, **PuBu**, **BuPu**, **RdPu**, **PuRd**

a. Three-class specifications

#	YlGn	YlGnBu	GnBu	BuGn	PuBuGn	PuBu	BuPu	RdPu	PuRd
1	247 252 185 / 3 0 27 0	237 248 177 / 7 0 30 0	224 243 219 / 12 0 12 0	229 245 249 / 10 0 0 0	236 226 240 / 7 9 0 0	236 231 242 / 7 0 0 0	224 236 244 / 12 3 0 0	253 224 221 / 0 12 8 0	231 225 239 / 9 9 0 0
2	173 221 142 / 32 0 43 0	127 205 187 / 50 0 20 0	168 221 181 / 34 0 25 0	153 216 201 / 40 0 15 0	166 189 219 / 35 15 0 0	166 189 219 / 35 0 0 0	158 188 218 / 38 14 0 0	250 159 181 / 0 38 12 0	201 148 199 / 20 38 0 0
3	49 163 84 / 81 0 76 0	44 127 184 / 85 27 0 0	67 162 202 / 75 12 0 0	44 162 95 / 83 0 70 0	28 144 153 / 90 12 27 0	43 140 190 / 85 0 0 0	136 86 167 / 47 60 0 0	197 27 138 / 20 90 0 0	221 28 119 / 10 90 15 0

b. Four-class specifications

#	YlGn	YlGnBu	GnBu	BuGn	PuBuGn	PuBu	BuPu	RdPu	PuRd
1	255 255 204 / 0 0 20 0	255 255 204 / 0 0 20 0	240 249 232 / 6 0 8 0	237 248 251 / 7 0 0 0	246 239 247 / 3 5 0 0	241 238 246 / 5 5 0 0	237 248 251 / 7 0 0 0	254 235 226 / 0 8 8 0	241 238 246 / 5 5 0 0
2	194 230 153 / 24 0 39 0	161 218 180 / 37 0 25 0	186 228 188 / 27 0 23 0	178 226 226 / 30 0 5 0	189 201 225 / 26 13 0 0	189 201 225 / 26 13 0 0	179 205 227 / 30 10 0 0	251 180 185 / 0 30 15 0	215 181 216 / 15 25 0 0
3	120 198 121 / 53 0 53 0	65 182 196 / 75 0 10 0	123 204 196 / 52 0 15 0	102 194 164 / 60 0 30 0	103 169 207 / 60 15 0 0	116 169 207 / 55 17 0 0	140 150 198 / 45 30 0 0	247 104 161 / 0 60 10 0	223 101 176 / 10 60 0 0
4	35 132 67 / 87 10 83 0	34 94 168 / 90 45 0 0	43 140 190 / 80 20 0 0	35 139 69 / 87 10 83 0	5 129 138 / 100 15 35 0	5 112 176 / 100 30 0 0	136 65 157 / 47 70 0 0	174 1 126 / 30 100 0 0	206 18 86 / 17 95 35 0

c. Five-class specifications

#	YlGn	YlGnBu	GnBu	BuGn	PuBuGn	PuBu	BuPu	RdPu	PuRd
1	255 255 204 / 0 0 20 0	255 255 204 / 0 0 20 0	240 249 232 / 6 0 8 0	237 248 251 / 7 0 0 0	246 239 247 / 3 5 0 0	241 238 246 / 5 5 0 0	237 248 251 / 7 0 0 0	254 235 226 / 0 8 8 0	241 238 246 / 5 5 0 0
2	194 230 153 / 24 0 39 0	161 218 180 / 37 0 25 0	186 228 188 / 27 0 23 0	178 226 226 / 30 0 5 0	189 201 225 / 26 13 0 0	189 201 225 / 26 13 0 0	179 205 227 / 30 10 0 0	251 180 185 / 0 30 15 0	215 181 216 / 15 25 0 0
3	120 198 121 / 53 0 53 0	65 182 196 / 75 0 10 0	123 204 196 / 52 0 15 0	102 194 164 / 60 0 30 0	103 169 207 / 60 15 0 0	116 169 207 / 55 17 0 0	140 150 198 / 45 30 0 0	247 104 161 / 0 60 10 0	223 101 176 / 10 60 0 0
4	49 163 84 / 81 0 76 0	44 127 184 / 85 27 0 0	67 162 202 / 75 12 0 0	44 162 95 / 83 0 70 0	28 144 153 / 90 12 27 0	43 140 190 / 85 0 0 0	136 86 167 / 47 60 0 0	197 27 138 / 20 90 0 0	221 28 119 / 10 90 15 0
5	0 104 55 / 100 25 90 0	37 52 148 / 90 70 0 0	8 104 172 / 100 35 0 0	0 109 44 / 100 20 100 0	1 109 89 / 100 25 65 0	4 90 141 / 100 30 0 20	129 15 124 / 47 95 0 5	122 1 119 / 50 100 0 0	152 0 67 / 40 47 0 0

RGB and CMYK specifications for nine sequential schemes with hue transitions in ColorBrewer, continued

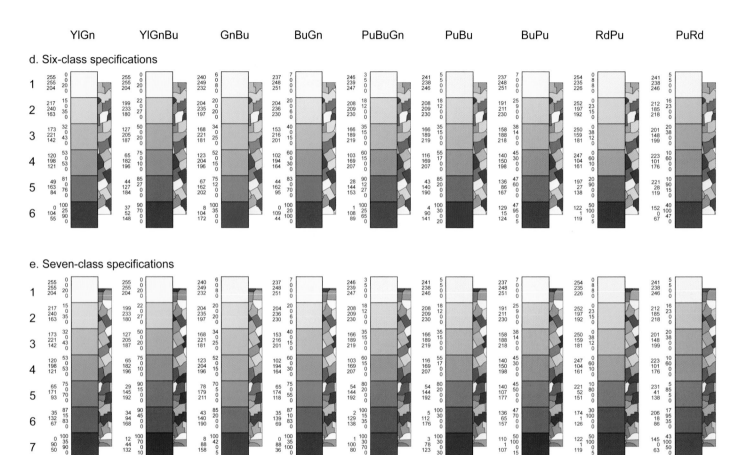

RGB and CMYK specifications for nine sequential schemes with hue transitions in ColorBrewer, continued

YlGn YlGnBu GnBu BuGn PuBuGn PuBu BuPu RdPu PuRd

f. Eight-class specifications

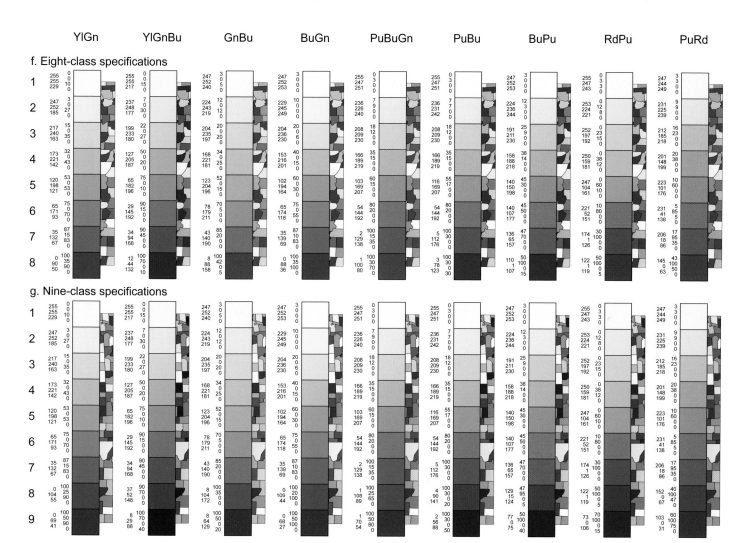

g. Nine-class specifications

RGB and CMYK specifications for additional sequential schemes in ColorBrewer (three with hue transitions and five single-hue schemes)

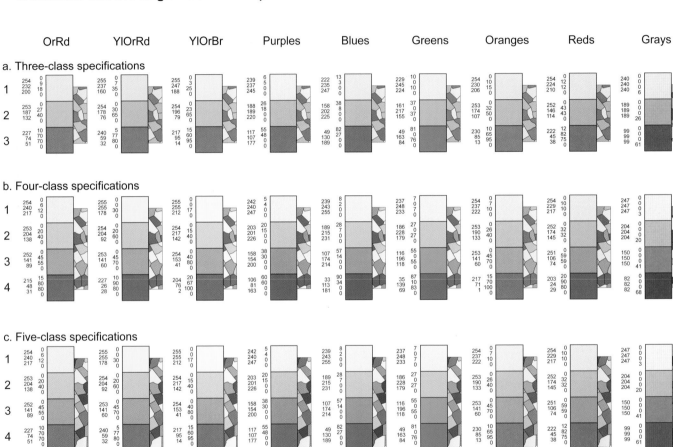

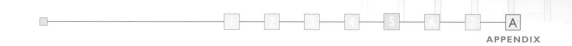

RGB and CMYK specifications for additional sequential schemes in ColorBrewer (three with hue transitions and five single-hue schemes), continued

OrRd   YlOrRd   YlOrBr   Purples   Blues   Greens   Oranges   Reds   Grays

d. Six-class specifications

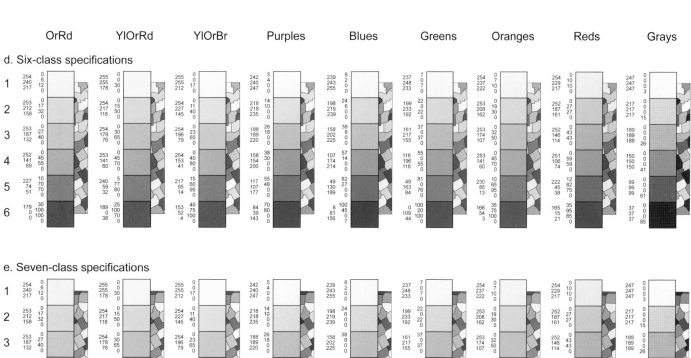

e. Seven-class specifications

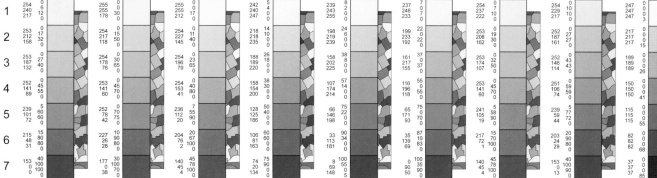

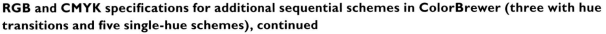

RGB and CMYK specifications for additional sequential schemes in ColorBrewer (three with hue transitions and five single-hue schemes), continued

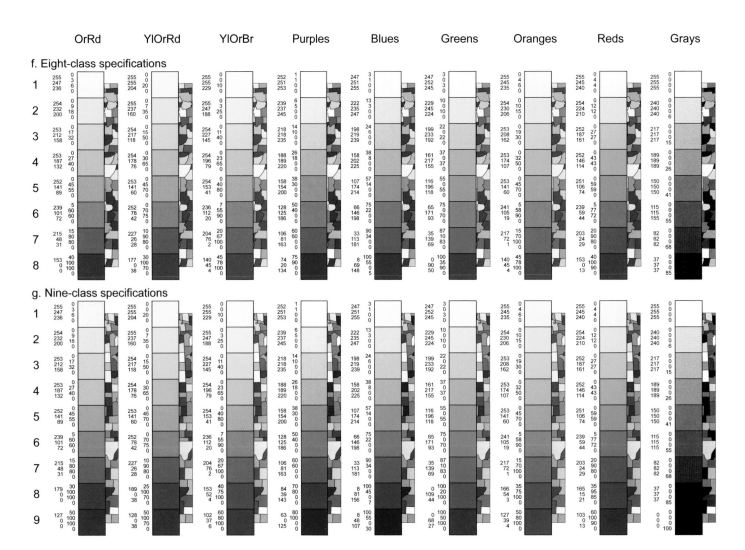

OrRd   YlOrRd   YlOrBr   Purples   Blues   Greens   Oranges   Reds   Grays

f. Eight-class specifications

g. Nine-class specifications

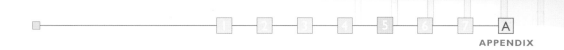
RGB and CMYK specifications for diverging schemes in ColorBrewer

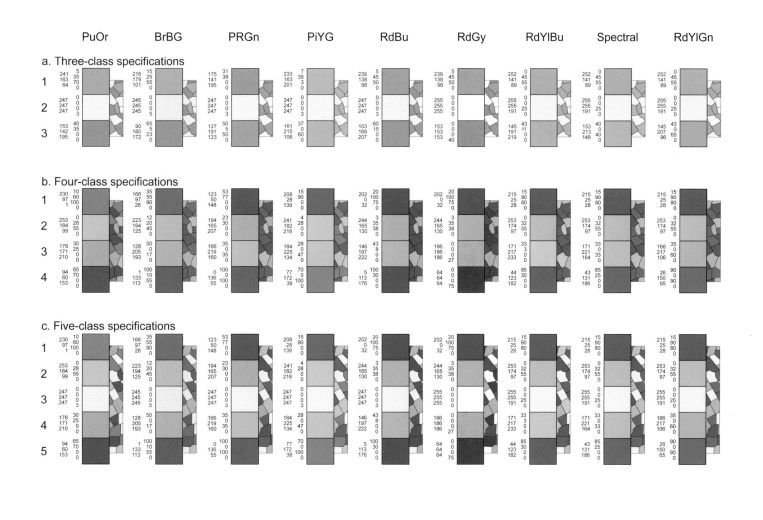

RGB and CMYK specifications for diverging schemes in ColorBrewer, continued

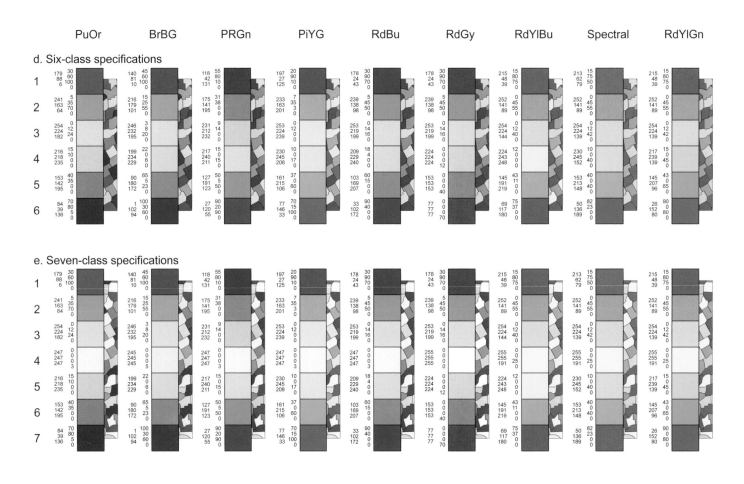

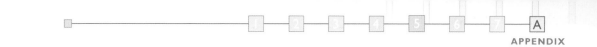
RGB and CMYK specifications for diverging schemes in ColorBrewer, continued

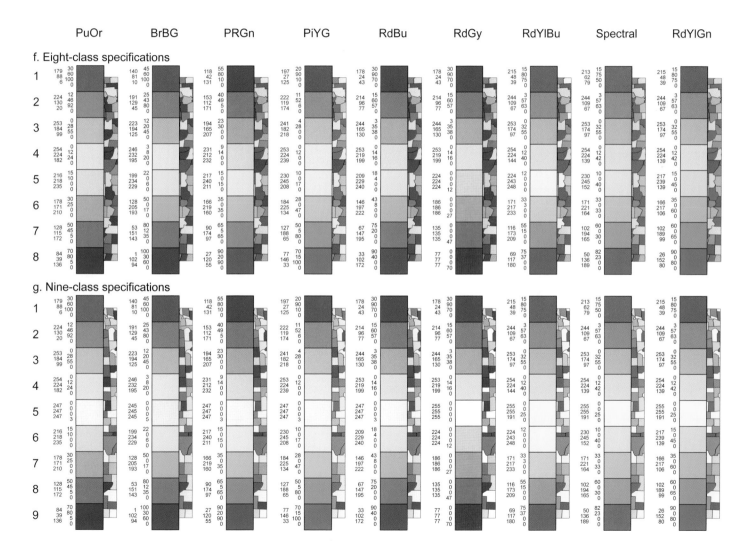

RGB and CMYK specifications for diverging schemes in ColorBrewer, continued

	PuOr	BrBG	PRGn	PiYG	RdBu	RdGy	RdYlBu	Spectral	RdYlGn

h. Ten-class specifications

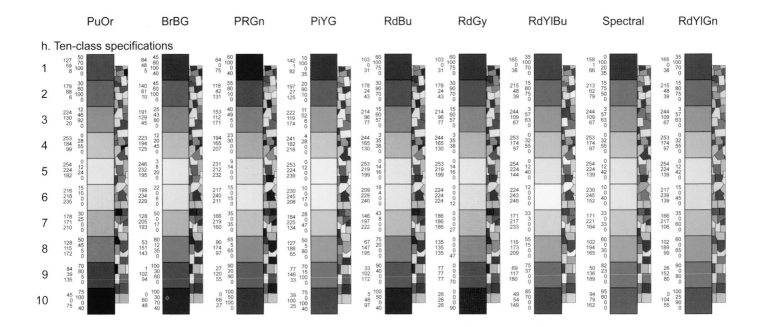

RGB and CMYK specifications for diverging schemes in ColorBrewer, continued

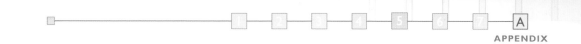

PuOr   BrBG   PRGn   PiYG   RdBu   RdGy   RdYlBu   Spectral   RdYlGn

i. Eleven-class specifications

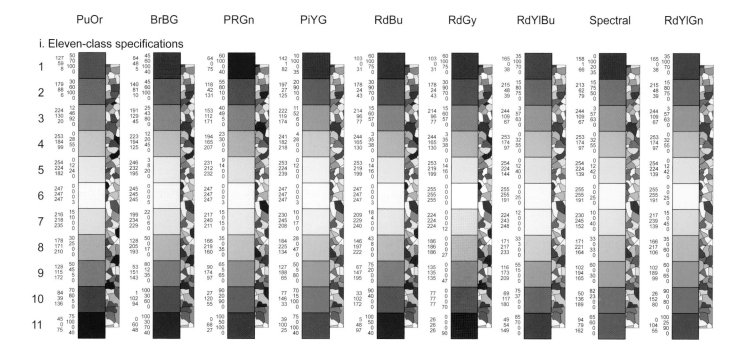

RGB and CMYK specifications for qualitative schemes in ColorBrewer

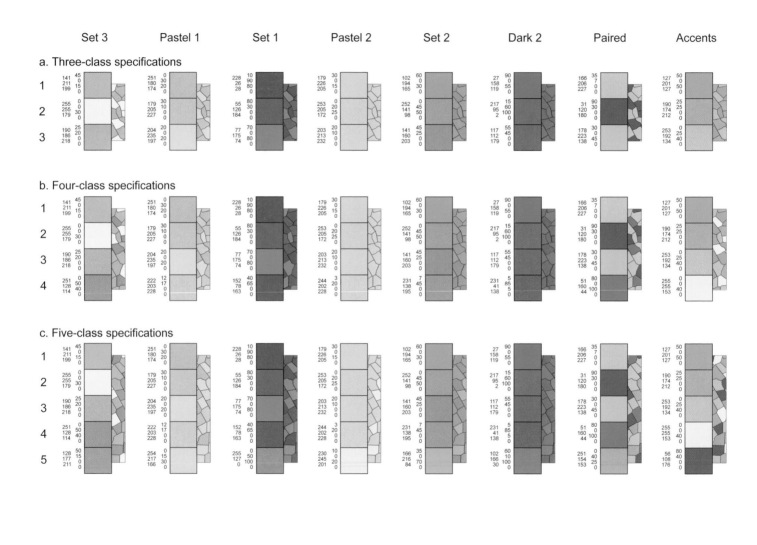

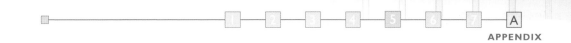

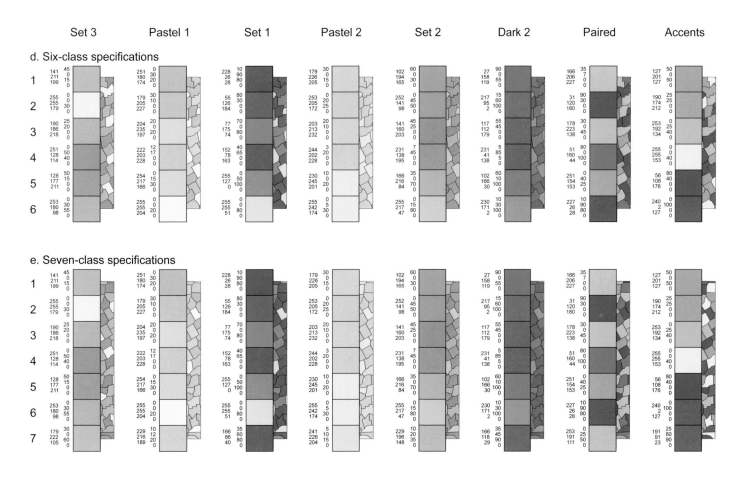

RGB and CMYK specifications for qualitative schemes in ColorBrewer, continued

RGB and CMYK specifications for qualitative schemes in ColorBrewer, continued

Set 3   Pastel 1   Set 1   Pastel 2   Set 2   Dark 2   Paired   Accents

f. All qualitative specifications

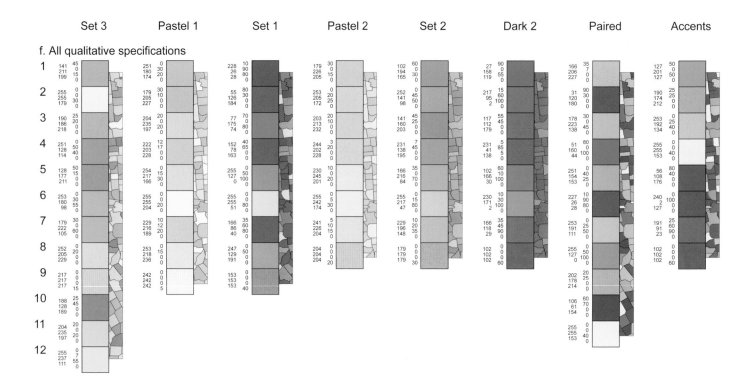

RGB and CMYK specifications for color-blind map readers

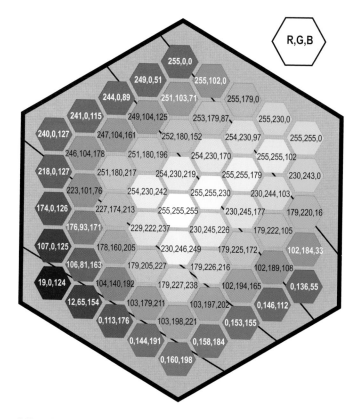

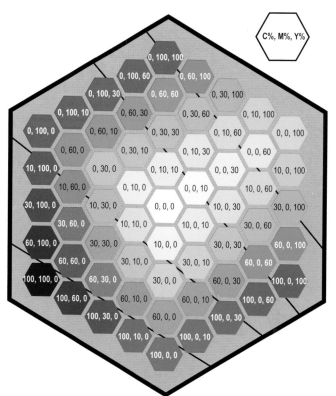

RGB combinations for color in the color-blind confusion diagram. Use these numbers in ArcMap for maps that will be viewed onscreen; colors entered as CMY will look peculiar onscreen in ArcMap, though it prints them well.

Combinations of 0, 10, 30, 60, and 100 percent of cyan, magenta, and yellow inks. Use these colors to prepare print products.

To use either of these diagrams, choose colors that are two or more zones apart when they have similar lightness.

FURTHER RESOURCES

Cartography books

Brown, Allan, and Wim Feringa. 2003. *Colour basics for GIS users.* Harlow, England; New York: Prentice Hall.

Campbell, John. 2001. *Map use and analysis.* 4th ed. Boston: McGraw-Hill.

Dent, Borden D. 1999. Cartography: Thematic map design. 5th ed. Boston: WCB/McGraw-Hill.

Dorling, Daniel, and David Fairbairn. 1997. *Mapping: Ways of representing the world.* Harlow, England: Prentice Hall.

Fairchild, Mark D. 2005. *Color appearance models.* 2nd ed. Chichester, West Sussex, England; Hoboken N.J.: John Wiley and Sons, Inc.

Kraak, Menno-Jan, and Ferjan Ormeling. 2003. *Cartography: Visualization of geospatial data.* 2nd ed. Harlow, England; New York: Prentice Hall.

MacEachren, Alan M. 1995. *How maps work: Representation, visualization, and design.* New York: Guilford Press.

Monmonier, Mark S. 1996. *How to lie with maps.* 2nd ed. Chicago: University of Chicago Press.

Monmonier, Mark S., 1993. *Mapping it out: Expository cartography for the humanities and social sciences.* Chicago: University of Chicago Press.

Muehrcke, Phillip C., Juliana O. Muehrcke, and A. Jon Kimerling. 2001. *Map use: Reading, analysis, interpretation.* 4th ed. (revised). Madison, Wisc.: J.P. Publications.

Robinson, Arthur H., Joel L. Morrison, Phillip C. Muehrcke, A. Jon Kimerling, and Stephen C. Guptill. 1995. *Elements of cartography.* 6th ed. New York: John Wiley and Sons, Inc.

Slocum, Terry A., Robert B. McMaster, Fritz C. Kessler, and Hugh H. Howard. 2004. *Thematic cartography and geographic visualization.* 2nd ed. Upper Saddle River, N.J.: Prentice Hall.

Cartography journals

Cartography research is published by journals in a variety of fields, including geography, GIS, and computer graphics. The following is a list of the primary cartographic journals.

Cartographica is published by the Canadian Cartographic Association. *www.geog.ubc.ca/cca*

Cartographic Perspectives is published by the North American Cartographic Information Society (NACIS). *www.nacis.org*

Cartography and Geographic Information Science is published by the Cartography and Geographic Information Society. *www.acsm.net/cagis*

The Cartographic Journal is published by the British Cartographic Society. *www.cartography.org.uk*

Cartography coursework

Content in this book is complemented with ArcMap exercises in the ESRI Virtual Campus course *Cartographic Design Using ArcGIS 9* by Cynthia Brewer.

Since 1998, ESRI Virtual Campus has offered GIS education and training over the Internet. Today, there are more than 250,000 Virtual Campus members located around the world, working in a wide variety of industries. Web-based training combines interactive exercises, conceptual material, and instructional resources to create a rich learning environment. Web courses provide self-paced instruction covering a variety of topics related to ESRI software, the theory underlying GIS technology, and the application of GIS tools to find solutions in particular fields.